ETHICS AND TECHNOLOGY

ETHICS and TECHNOLOGY

Innovation and Transformation in Community Contexts

John Hart

The Pilgrim Press / Cleveland, Ohio

The Pilgrim Press, Cleveland, Ohio 44115

© 1994, 1997 by John Hart

First Pilgrim Press Edition, 1997
Originally published by the William C. Norris Institute,
Bloomington, Minnesota, 1995

All rights reserved
Printed in the United States of America on acid-free paper

02 01 00 99 98 97   5 4 3 2 1

Library of Congress Cataloging-in-Publication Data
  Hart, John, 1943–
    Ethics and technology : innovation and transformation in community contexts / John Hart.
      p. cm.
    Includes bibliographical references.
    ISBN 0-8298-1222-9 (alk. paper)
    1. Social ethics. 2. Technology—Social aspects. 3. Technology—Moral and ethical aspects. 4. Technological innovations—Social aspects. I. Title.
HM216.H245   1997
303.48'3—dc21
                                        97-30708
                                            CIP

For William C. Norris:

technology innovator, computer pioneer, international entrepreneur, creative education proponent, social change advocate;

recipient of the Gold Mercury Award for Contributions to International Relations and World Peace (USSR, 1980), recipient of the National Medal of Technology (USA, 1986), and of numerous other honors;

mentor and friend

CONTENTS

Preface ix
Acknowledgments xiii
Introduction: Ethics and Technology xiv

1 Ethics and Development *1*
2 Individual and Society *31*
3 Employee and Employer *47*
4 Corporation and Community *70*
5 Entrepreneurship, Employment, and Environment *90*
6 Spirit, Science, and Society *110*
7 Tradition, Technology, and Transformation *141*

Appendix A: Case Studies *151*
Appendix B: Reflection Questions *167*
Appendix C: Projects *171*
Selected Bibliography *173*

PREFACE

The announcement in February 1997 that a sheep had been cloned at the Roslin Institute near Edinburgh, Scotland, raised important questions about the relationship between ethics and technology. Human responsibility was linked to scientific capability in a new way because the sheep cloning opened the possibility of human cloning. The event made people wonder anew: Just because technological feats are possible, should science carry them out? Does *can* equal *ought* in the quest for human progress?

In what might appear to be an unrelated event, Pope John Paul II restated in Rome a month later a Christian social teaching that has been particularly highlighted in the twentieth century: that people take precedence over profits. Addressing workers in Rome, the pope reminded "those tempted to affirm the predominance of technique, reducing people to 'commodities' or instruments of production, that the proper subject of work remains the human person."

Although John Paul in Italy was advocating the inherent dignity and worth of the worker in the production process, his thoughts related well to the events in Scotland. As news spread about the cloning, a long-hidden ethical debate surfaced for public consideration: Should scientists clone humans? Thus, John Paul's concern—that technique in the production process in industry is given greater emphasis than the contributions of workers—might be linked to a concern that technique in biotechnology enterprises will be given priority over the inherent dignity of human persons. How would the ethical ideal that people have priority over profits play in such a social arena?

In thinking about this last question, one might do well to consult the literary works produced during the past century and a half that speculated about technological development and its impact on society. New questions might emerge as a result of such reflection. For example, will Aldous Huxley prove to have been a visionary for the field of genetics with his novel *Brave New World* as Jules Verne was a visionary about space travel with *From the Earth to the Moon*, and about submarine travel with *Twenty Thousand Leagues under the Sea*? Will Michael Crichton's warnings about the dehumanization of humans in the harvesting of their body parts in *Coma* and about the perils of genetic manipulation in *Jurassic Park* be lost in the glitz of movie special effects? Will Herbert Spencer's social Darwinism and B. F. Skinner's behavioralism be resurrected and linked to human cloning possibilities in order to justify the creation of slaves and soldiers? Will cloned human bodies, like the technologically sustained brain-dead bodies in *Coma*, be regarded as producers of body parts rather than as human beings? What practices against people — especially against marginalized peoples — might be justified? In the initial phase of the Spanish conquest of the Americas, native peoples were called "beasts who talked" rather than human beings in order to justify their enslavement; Spain had already outlawed slavery. Will a pope feel obliged to issue a statement declaring that the cloned new humans have souls, which would parallel the 1537 document *Sublimis Deus*, issued during the Spanish conquest by Pope Paul III, which declared that native peoples have souls and are human?

Biotechnology will continue to have positive uses. Disease remediation and pollution reduction come to mind in this regard. But the twentieth century has seen the Holocaust in Germany, apartheid in South Africa, and the continuation of genocide against indigenous peoples throughout the Americas. How will people regarded as inferior fare if ethics does not keep pace with technological development, or if ethics is rejected as irrelevant to that development?

People engaged in discussions about cutting-edge science and technology projections sometimes pretend that such exchanges are merely intellectual exercises, abstract considerations of present and projected techniques and their potential for profit and so-called progress. What is missing in these discussions is consideration of the social context of human enterprise. What is missing, too, is a focus on the potential of technological innovation to have a negative or a positive transformative impact on society and on the earth community in which society is situated. In issues of biotechnology, such as cloning, for example — and in related issues such as gene mapping and gene patenting — several important questions (especially those that focus on the priority of needs over wants) are sometimes brushed aside:

- Is there a need for this process or product, or is its purpose primarily profit?

- Can the social or environmental need be met solely by this process or product?
- What will be the social and environmental impacts of this process or product?
- Will earth and its biotic community benefit from this process or product?

Over the past three decades of my professional life, I have been involved with distinct and diverse groups of people who have been the architects or the recipients of technological change. I have worked with farmers and farmworkers, with business leaders and labor leaders, with human rights and civil rights organizations, with political and religious leaders, with teachers and students, and with members of oppressed economic classes and ethnic groups. I have noticed at times in this diversity of people a disjunction between the theoretical and the practical: everyday life seems to leave little room for reflection on the course of human events and on the changes in environmental conditions. In part, I believe, this is because the ordinary citizen lacks access to the world of technology and the world of ethics.

This book, then, opens a window on the world of ethical reflection, with a focus on the world of rapid technological change. To borrow a term from the computer world, this text attempts to be user friendly, blending theoretical consideration of ethics and ethical issues with real-life situations and decisionmaking in an interactive way. I have endeavored to provide the reader with some concretization of the abstract, particularly through the Content in Context features created to illustrate ethical–technical interactions.

Ethics and Technology is offered for consideration by a wide readership: concerned citizens, scientists and technologists, elected officials, students and teachers. Members of all these constituencies have some stake in the direction taken by technology in this era of rapid technological developments and extensive technological possibilities.

Ethics and Technology has been constructed to be helpful to people from a variety of backgrounds. The main body of the text discusses salient issues and offers ethical principles for consideration. Three appendixes are provided at the end of the book to assist readers who wish to explore more deeply the ethics-technology interaction. Appendix A describes organizations whose efforts in whole or in part embody the principles cited in the book; Appendix B offers questions crafted to stimulate further theorizing about issues described in the book; and Appendix C proposes community or campus projects designed to put ethics and theory into practice.

Numerous ethical issues related to technology are not discussed here. My focus is on issues of political and economic justice and care for creation as business and technology utilize the earth's resources to meet human needs and wants. I acknowledge my limitations of expertise and space and defer to others discussion of issues in such areas as biotechnology and medical technology. I do believe, however, that reflection on those topics might benefit from a consideration of ethical principles and practices suggested in this book related to communitarian ethics, environmental concerns, and economic issues, and the proposals for an ethics of transformation.

The future we bequeath to our children and to our descendants beyond them awaits our commitment today to integrate ethics and technology within a sustained earth community. Let us be creative in envisioning a sustainable future, and committed to making it a reality.

ACKNOWLEDGMENTS

Ethics and Technology began as a module for a computer-based course initiated by the William C. Norris Institute and focused on the management and utilization of technology. At the invitation of William Norris I began to develop the module, which was to be used to educate people in the United States and abroad. The written text grew as I considered various aspects of the relationship of ethics and technology and as William Norris suggested additional topics from the perspective of a business leader; thus it evolved from module to manuscript. I would like to thank Bill Norris for his invitation, his encouragement, his insights, his support, and his friendship, all of which combined to stimulate the creation of *Ethics and Technology*.

The time to write this book came from a sabbatical leave granted by my faculty colleagues at Carroll College, Helena, Montana; and through the loving encouragement of my wife, Jane, and our children, Shanti and Daniel. I thank them.

I also want to express my gratitude to the folks at The Pilgrim Press, who have masterfully completed the writing process by preparing the manuscript for publication and then guiding it through the process: Timothy Staveteig, Editor; Marjorie Pon, Managing Editor; and Ed Huddleston, Associate Managing Editor.

Finally, I would like to express my gratitude to my professors at Union Theological Seminary, New York, from which I graduated nearly twenty years ago, for their teaching and guidance in the area of ethics; to the native elders across the continent who have shared their insights and given instruction about their struggles and their spirituality, and have offered me the warmth of their friendship; to my faculty colleagues at Carroll College for stimulating conversations, outdoor adventures, and friendship, all of which helped to lighten a heavy workload; and to the social activists of diverse backgrounds and perspectives with whom I have worked and continue to work in concrete ways to create a just society and a renewed earth.

Thank you, one and all.

INTRODUCTION: ETHICS AND TECHNOLOGY

During the past century a dramatic increase in technological innovation has brought about extraordinary benefits for humanity. Through advances in medicine, aeronautics, genetics, and computer science, people can live better and longer, and work faster and more efficiently. Technology also has brought problems into human life. In the areas mentioned, for example, technology has produced mass-destruction weapons for biological and nuclear warfare, displaced people from employment through the shift from labor-intensive to capital-intensive production processes, and despoiled the earth through careless extraction of natural resources.

Technology is neither entirely a blessing nor entirely a curse for human societies. Technology will benefit or harm humanity according to the intentionality and integrity of technologists, the social consciousness and conscience of business and political leaders, and the awareness and activity of individual or organized citizens. The intentionality of technologists is the purpose they intend for their work, not only in the strictly technical sense of how they perceive the utility of what they are developing, but also in the volitional sense of how they want their work to be used. The integrity of technologists refers to their conscientious indignation and their courageous intervention when they foresee that what they have devised or developed, conceptualized or concretized, will have harmful impacts in the hands of others.

Business and political leaders have varying degrees of social consciousness—their awareness of social injustices and problems and their recognition of factors causing them. They also have varying degrees of social conscience—their sense of need to alleviate or eliminate those injustices and problems and their extent of resolve to do so. Leaders with a positively developed, community-oriented social consciousness and conscience analyze technology in terms of people then profits, or people and profits; those with a negative, individually or solely company-oriented social consciousness and conscience analyze technology in terms of profits then people. Because of their positions of power, leaders for better or worse can direct or redirect technologists' intentionality or productivity.

Citizens respond to technology when they become aware of the potential or actual implications of a present or proposed technology for people and for the planet; and then become active in relating openly to business and government in order to promote beneficial technology and resist harmful technology. Citizen participation in the politics of technology development can offset the antisocial use of power by business and political leaders, stimulate business to produce a socially and environmentally more benign and better product, help business to find the most appropriate siting for its manufacturing operations, and facilitate overall the establishment of a collaborative process for creating good jobs by assisting with the start-up of technology-based companies that will be community-responsive and environmentally responsible. Citizen awareness and activity should work to prevent community harm and also to promote community benefit, including, where possible, related company benefit: citizens and communities need goods and employment, and they share responsibility for the impacts (social and environmental) of the goods they demand as individuals for consumption and of the process that produces those goods. Communities are responsible, too, for the economic well-being of their individual members and the related social stability of the community as a whole.

The technological revolution is a potential boon for individuals, nations, and the international community. Whether or not that potential will be realized will depend on the values present in the development, management, and utilization of technology. Over the millennia of recorded human history, social values have emerged at times from sacred and secular (or humanist) traditions. Ethical considerations for technological development draw both from the riches of those intellectual resources and from the experiential resources of people involved in the concrete practice of industrial production.

Ethical considerations for the innovative process—from the formulation of a creative idea through its concretization in production, and the utilization and eventual disposal of the resulting product or its remnant components by the manufacturer and/or by those who acquire it—focus on the interplay between the individual and society (the balance between private and public good); the working relationship between employer and employee; the corporate impact on (local, national, and international) communities; and the effect of technologies on people and planet.

The term *business* as used in this book ordinarily refers to an industrial corporation, because the book focuses on technology, but the ideas presented usually would be applicable to other types of business as well. Fundamental social ethical insights are presented and related to technological innovation. The presentation of these insights and their relation to technology does not include consideration of all the intricacies of ethical analysis. It does, however, provide basic intellectual and ethical tools for a creative analysis and evaluation of the ethical implications and social impacts of technology: the existing technology and proposed technology of the present, and the as yet unimagined technological possibilities of the future.

The ethical principles are formulated to promote individual and corporate responsibility for people and planet. They are meant to be challenging and to provoke consideration of what social, commercial, political, and environmental changes might occur locally, regionally, and globally if they were to be accepted and implemented. Those who do not agree with some or all of these principles should reflect on the reasons for their disagreement, on what alternatives they would propose, on why they would want those particular alternatives, and on what might be the respective impacts of (*a*) the ethical principles proposed in the text and (*b*) the proposed alternative ethical principles.

Readers are encouraged to consider these principles from varying perspectives to get a sense of how others might view them: for example, a businessperson or technologist might approach them not as an entrepreneur, engineer, technician, or employee but as a community resident without a vested interest in the proposed development. As such a community resident, would they tend to favor the principles? If so, how might they reconcile how they would like to be treated as a concerned citizen with how they would like to be received as an entrepreneur or employee by the concerned citizens of a community in which they want to locate their business? Similarly, a community resident or environmentalist might try to consider the principles from the perspective of a business leader or worker; an employer might view principles as an employee would; an employee as an employer; and so forth.

Those who become proficient at this kind of role-shifting will have the potential to prevent problems from arising and to resolve problems more easily when they have arisen. They will also be better prepared to adapt to changing circumstances resulting from the environmental impacts of production, new government regulations, new political leadership with existing regulations, citizen concerns, resource shifts, production or product obsolescence, employee changes, shareholder pressures, and so forth. When people are able to understand others' perspectives on the consequences of human actions, cooperation can replace conflict in society.

CHAPTER ONE

Ethics and Development

Developing new goods or services to meet human needs or wants requires innovative ideas, and the means to realize them in concrete production processes. The impacts of human creativity and productivity result not only from the accuracy of projections about such impacts, but also from the values brought to innovation and development by researchers, developers, entrepreneurs, employers and employees, and by the governing officials of the political entity or entities in which they fulfill their respective responsibilities.

In the areas of research, development, production, distribution, and disposal, then, ethics has an important role to play in the process from idea to implementation, and in company–community relations. Ethical considerations can make the difference in whether a particular innovation proves to be helpful or harmful to a specific society, to the broader human community, and to the regional or global environment.

The role of ethics in the business and scientific worlds is accepted when it is seen as an aid to thoughtful and responsible research, development, production, and distribution activities. It is misunderstood when businesspeople or scientists view ethics only as an unrealistic constraint upon legitimate and even moral business activity. It is denied when businesspeople or scientists declare that their development is "value free" or "value neutral," as if that could ever really be the case. It is rejected when businesspeople or scientists realize that they are engaged in legally or morally questionable activities and resist constraints upon those activities.

Ethics challenges "business as usual" in all these cases. Those who appreciate the contribution ethics can make might find themselves challenged to think in new directions and to act in alternative ways in order to respond positively to ethical values bearing on their enterprise. Those who are wary of the intervention of ethical ideas into a business or scientific context because they feel legitimate professional practices are sufficient fail to realize that ethics challenges all parties in that context to fulfill their responsibilities: production personnel and general public, industry and government, employer and employee, and so forth. For example, ethics advocates

> the rights of labor (e.g., to a just wage and the right to strike) + the responsibilities of labor (e.g., to do good work and to properly care for plant facilities);
>
> the rights of business (e.g., to a just return on investment and to good work by labor) + the responsibilities of business (e.g., to pay just wages, provide safe working conditions, and market a good product);
>
> the rights of government (e.g., to protect the common good of the community and to promote just wages) + the responsibilities of government (e.g., to establish an equitable tax structure and perhaps to give rebates to small companies to allow their reasonable compliance with minimum wage, safety, and benefit requirements imposed by government); and
>
> the rights of citizens (to good and safe products and to a healthy environment for their community) + the responsibilities of citizens (to pay in a timely manner for goods received and to assist in the safe disposal of unwanted byproducts resulting from the production of goods required by consumers).

In all of these efforts, ethics promotes responsible interaction among all the parties involved in or benefiting from productive business activity. (The just-wage social value, which will be discussed in more depth later, essentially means that employees are entitled to compensation that enables them to meet their basic individual and family needs, including food, clothing, shelter, energy, and medical costs.)

Those who claim that ethics is irrelevant to their particular situation, on the grounds that no values are present and no value judgments are pertinent, blind themselves to the very human reality that subjective biases and value choices of one type or another are inherent in all human activity, and that ethics in that context can help

CONTENT IN CONTEXT 1.1

Value Neutral?

Dr. Robert Heller is a nuclear physicist working on a missile warhead that will carry a biological payload to enemy targets. His country's leaders declare "freedom" to be a global ideal. His innovative idea has been to irradiate the virus so that even when it does not kill its intended victims outright, it will render them (human, animal, or plant) sterile; when it does kill life forms, it contaminates the soil around them. Dr. Heller is pleased with his innovation: although it has the potential to cause irreversible damage to people and to the area in which it lands, he hopes that its mere existence will serve as a deterrent to enemy attack, or even to any foreign power's disagreement with his nation's policies. After all, his country has good values, and others should respect its leadership.

In an exuberant moment at dinner one evening, Dr. Heller mentioned his creative work to his wife, Alma, a hospice nurse. When Alma wondered aloud about the propriety of such a weapon he replied: "My research is value neutral. I'm just putting my intellect at the disposal of my country. I don't decide to use the warhead, the politicians do; I don't launch the missile, the military does. If the horror of this weapon is unleashed upon others, it is their responsibility, not mine." Alma responded: "There would be no horror without your creativity. Besides, Thoreau once wondered why everyone has a conscience, if they're supposed to resign it to their government."

Can research be "value neutral" when it might result in harm to people, other life forms, or the earth? Should people involved in research and development question the potential impacts of the results of their efforts? Should they project potential dangers from a change in government or from irresponsible ownership or management of their company, and reject harmful projects?

them evaluate seriously their rationale for their choices and the implications of those choices.

In the area of weapons development, for example, some scientists and manufacturers declare that since they do not personally decide whether or not to use their developing or developed biological or nuclear weapon and do not personally launch the weapons, they are not responsible for what such weapons do to people or planet. Practically speaking and politically speaking, of course, they must know that their leaders might act irresponsibly with their innovation, wreaking horrendous suffering and irrevocable harm. Those who devise or develop such weapons, then, will be the remote cause of their consequences, and should weigh beforehand basic ethical values (such as compassion and love of neighbor) and the potential devastating consequences of the fruits of their labors. Of course, research scientists and businesspeople involved in innovative explorations for military or commercial purposes cannot be expected to realize all or, sometimes, even some of the harmful uses of what they develop and produce. As they do come to that realization, however, they will have to make serious and difficult ethical choices as to whether they wish to continue to be involved with a specific project and even whether the public should be informed about what is under development: they must balance allegiance to the chain of command with allegiance to the broader society. A general or a president might order one course of action, but conscience or fear of retribution might order another. The Nuremberg trials after World War II demonstrated, at least in the abstract, that people believe individuals should act morally even when fearful for personal safety, or accept the consequences if they do not. Unfortunately, usually only the losers of conflicts are brought to trial for war crimes; atrocities committed by the winners rarely bring retribution.

Those who seek to exclude ethics from their activities because ethical principles conflict with their privately profitable but socially harmful production will find that concerned others will question their activities, suggest ethical alternatives, and, if necessary, seek to impose legal constraints upon those activities. In a situation where unscrupulous or irresponsible conduct momentarily prevails and moral suasion fails, legal sanctions might be sought by those concerned about the public (or planetary) good. For example, in a case of environmental pollution, plant officials initially might deny that smoke emitted (or effluent released, or waste shipped or buried) is harmful; then when toxins are discovered say that their levels are "acceptable" by industry or government standards; then fight investigation as unjustified intervention in commerce and as job threatening; then lobby to change existing environmental laws and to prevent enactment of new environmental laws—all this rather than produce a safe product in a safe way, and rather than ensure public well-being in the pursuit of private profit.

In the industrialized nations, people often turn to material production processes and possessions to satisfy both basic needs and created or developed wants. In the area of needs, food, clothing, shelter, medical care, and energy require material resources, while psychological, social, and spiritual satisfaction do not, but people seek material things to meet all these needs. In the area of wants, material resources are preeminently required. The major consequences of this focus on material consumption are that sufficient goods are lacking to meet some people's present and future needs because those goods are being used for other people's present wants; and that the earth is being increasingly exploited and stripped of its resources for the short-term satisfaction of the material wants of those few, imperiling the long-term sustainability of all people, of other life forms, and of the earth itself.

GROWTH AND PROGRESS

Some people define economic growth and progress solely in terms of increased utilization of material resources and increased expansion of production processes; the carrying capacity of the planet in terms not only of population but also of resource availability and use is considered only superficially, if at all. These people see a lack of "progress" in growth that includes not only use of new resources but reuse of old materials; job creation and expansion in new intermediate or labor-intensive technologies; or increases in the number of innovative small businesses rather than or in addition to expansion by large corporations. An additional problem is that some business owners and managers believe that business growth and prosperity are to be found in a focus on "wants": they do not realize that "needs"—in terms of goods, services, and jobs—can also present profitable business opportunities and spur innovation and growth while enhancing community well-being.

A new understanding of "growth" is needed if sustainable economic development, employment provision, and environmental conservation are to become accepted as values and implemented as practices in local, regional, and national communities. Economic development can be sustained if realistic assessments are made about the public's need for a product; about the availability of resources (natural, recycled, energy, and human) to manufacture the product; and about the vital role that small companies play in the introduction of new products. Employment provision can be sustained if companies accept a reasonable return on investment rather than constantly seeking to maximize profits through downsizing. Environmental conservation can be sustained if economic development and employment provision take into account the need to harvest or extract resources in a way and at a pace that respects local ecologies while meeting human needs.

ETHICS

Ethics can help to redefine notions of growth and thereby enhance the well-being of both company and community. The discipline of ethics suggests consideration of fundamental human values in the development and utilization of technology to meet human needs and wants. Ethics focuses on individuals and societies, seeking to balance such issues or concerns as personal wants and societal needs, personal needs and societal wants, competing needs, and competing wants.

Ethics might be defined as a systematized consideration of appropriate conduct, based on values expressed in principles, for theoretical and actual situations. Ethics is systematized in that it is an organized reflection on individual and social moral issues and on human responses to such issues. It considers appropriate conduct for individuals and groups based on generally shared values (such as respect for life, respect for property, and respect for human dignity). It states those values in principles or guidelines for conduct which are applied to or at least considered in general or particular cases, either projecting theoretical circumstances or focusing on concrete circumstances. It seeks to resolve conflicts between competing principles or claims in present or projected contexts, and to allow for exceptions to one principle when another principle appears to have priority in a particular situation.

Normative Ethics and Contextual Ethics

There are two basic types of ethical systems: normative and contextual. Normative systems focus on principles of conduct deemed to be ordinarily operative in all situations. Exceptions to those principles are allowed, after careful consideration, in unique or unusually complex circumstances, particularly when principles are in conflict. Contextual systems focus on the exigencies of the situation. Each situation is seen as unique, having evaluative elements in itself; principles are exceptional, few in number if existing at all, and serve only as general guides. In contextual systems the projected consequences of human action or inaction serve to focus decisionmaking.

> normative ethics: principles for right conduct
> contextual ethics: situational decisionmaking

<u>Each system has its benefits and drawbacks.</u> Normative systems offer helpful guidelines but could become very rigid and allow no (or very few) exceptions to established principles; contextual systems offer more flexibility but might degenerate into moral relativism, where each possibility is given equal weight in a given situation. On a societal level, what might happen is that normatively oriented people might try to impose their standards on others, while contextually oriented people might deny any objective standard and pursue individualistic interests over social benefits to the commonweal.

The result might be dictatorship or anarchy, depending on which group prevails. In normatively oriented cultures the individual could become subordinate to the group or to those members of the group currently in control; a strong ruler might be sought to uphold the group's values; divine authority might be claimed for the ruler(s), resulting in a "theocracy." In contextually oriented cultures each person could become an authority in him- or herself, and might decide appropriate situational conduct on the basis of selfishness. When several people in such a situation all express their individual autonomy, rather than seeking more objective guidelines or even alternative viewpoints, conflict results. Fistfights, brawls, or wars might then occur, depending on the number and type of inter-

CONTENT IN CONTEXT 1.2

A Nation under God

Rev. Elmer Martro, while watching the television news one evening, thought over the possibility of running for political office because of the moral decay evident in U.S. society. The traditional Christian values which he thought should permeate the national agenda were being set aside in the name of religious "pluralism," sexual "parity," economic "prosperity," social "progress," and world "peace." Society was falling apart, pushed to immorality because of the trash that passed for network television programming, and because children couldn't pray in school anymore.

Rev. Martro wanted the United States to be a Christian nation "under God," as stated in the Pledge of Allegiance, and to trust in God as engraved on U.S. currency. He believed that men should be the undisputed heads of households and that wives and mothers should not be out working. He wanted to lower taxes, eliminate welfare, and build more prisons. He was against gun control but for stiffer penalties for criminals. He wanted prayer in public schools and birth-control instruction out of public schools. And he wanted decreased "appeasement" and increased military spending in order to confront the threat of "atheistic communism."

Rev. Martro spent a night in prayer debating his entry into the political arena. Although he had no vision, nor did he hear a heavenly voice, he confidently decided that God had chosen him to set his nation right. The alternative would be that the United States would not be guided and guarded by God.

What conflicts might result if Rev. Martro is successful in his quest? Would the conflicts differ if Rev. Martro were Rabbi Meyer in Israel or Mullah Mohammed in Iran? What regional and global conflicts might occur if all three were simultaneously successful in promoting a theocracy in their respective nations?

ested parties in the situation and the degree of rigidity they have about their position.

In some heterogeneous societies, a certain "moral pluralism" (as distinct from the "moral anarchy" probable in a society lacking some minimal number of generally accepted principles) might develop and be seen as appropriate, where such a situation does not harm either society as a whole or its individual members. In this case, respect exists for majority or minority perspectives at variance with one's own views, provided, of course, that the respect is mutual and does not lead to harm to oneself or to others with whom one is associated or with whom one is in sympathy. In such a context, for example, a citizen who is a member of the military might disagree with the pacifist stance of another citizen, but grant to that citizen the right to live according to his or her own beliefs; the pacifist, in turn, might acknowledge the soldier's belief that military service protects the nation but disagree with the soldier's military means of providing that protection.

The anarchical society and the dictatorial society both have major drawbacks: in the former, "survival of the fittest" might become the basis for human interaction; in the latter, the people would suffer from oppression by the individual or party abusing political and economic power at a given historical moment.

Social Ethics

A balance between normative and contextual modes of ethical thought would be a system of social ethics that (1) has a generally normative perspective but allows reasonable exceptions to norms in particular circumstances; (2) is oriented toward the common good but still respects individual rights; (3) seeks to balance or prioritize competing principles; and (4) strives equitably to resolve competing claims of individuals, competing claims between individuals and society, and competing claims between communities, in a way that benefits all to the greatest extent possible.

Social ethics:

- general principles with particular exceptions
- common good and individual rights
- balance between competing principles
- equitable resolution of competing claims

The role of social ethics in business and technological development, then, is to provide general principles of conduct that inform particular circumstances. These principles can promote consideration of the potential impacts on individuals, communities, and the environment of actual and proposed processes and products of development.

Communitarian Ethics

The focus of many proposals for ethical conduct in Western industrial societies—and in other societies under their influence or domination—is individual rights and responsibilities. In their origins, such approaches were a legitimate reaction to social beliefs, pressures, and laws that actually or apparently stifled individual dignity and rights. These approaches maintain a certain legitimacy in contemporary situations in which individual citizens lose their fundamental rights and are forced to march to the collective drum, whether that instrument is beat by a dictator from the political right or bureaucrats from within a nation whose leadership has merely substituted party leadership for leadership by a moneyed economic and political elite. The focus on individual rights, for all its appropriateness and advantages, is inherently flawed in our world when it is not accompanied by a corollary reminder of individual responsibilities and not related to community rights and responsibilities.

Communitarian ethics, by contrast, emphasizes the mutual responsibilities of the individual and society in concrete social circumstances. It recognizes the social nature of human beings and the responsibility of the state, when people gather into communities, to mediate citizens' differences and promote their common good.

In these pages communitarian ethics is advocated, and the position is advanced that communitarian ethics is more valid for human interaction than individualistic ethics. In the United States and much of western Europe, this is not a popular position to take, in part because of the individualism and greed promoted by laissez-faire capitalism and the economic, political, and media structures that promote it, and in part because of a legitimate reaction against dictatorial soviet collectivism pursued in the name of socialism and the good of the people. However, communitarian ethics is the approach most faithful to the Western intellectual and religious tradition and

CONTENT IN CONTEXT 1.3

Autonomy or Association?

Martha Smogge became very upset as she read the new company directive on smoking: one lounge with air purifiers had been set aside by management as the sole indoor smoking area. Martha fumed: "I have every right to smoke. It's my body to do with as I choose." She confided her outrage to her friend, Myrtle Lusher. Myrtle was a nonsmoker, but the day before at their two-martini lunch their waitress had calmly suggested to Myrtle, who was eight months pregnant, that she might want to consider the effects of alcohol on her future offspring. "I know what you mean," Myrtle sympathized. "It seems as if the government and individuals want to impose their values on us." Joan Ark, a co-worker allergic to smoke, whose desk shared a partition with Myrtle's, overheard the conversation and noted aloud to the partition that a judge had once observed that "Your right to extend your fist stops where my nose begins," and wondered if smoke and alcohol might be included in the analogy. "After all, you should have a little love for your neighbors. It's not just you who inhale the smoke or imbibe the alcohol." Martha left Myrtle's side with a loud snort of disagreement, and returned to her desk.

How autonomous are Martha and Myrtle, and how are they associated with their community? To what extent should we be conscious of the effects on the community of our courses of action? What rights does the community have to regulate individual conduct, and for whose benefit: the individual involved, or only other people affected by the individual's actions? Should individualism be tolerated when it harms others: tobacco that produces secondhand smoke injurious to nonsmokers; drunk driving that may cause an accident that injures or kills not only the drunk driver but other people?

also most consonant with scientific understandings of the internal operations of organisms, the interaction of organisms, and the relation of organisms to the earth they inhabit.

Communitarian perspectives guided early peoples around the world as they cooperated in the hunt, for defense, and later in agriculture. This kind of cooperation is still evident today in preindustrial cultures living in rainforests, on mountain heights, and over plains expanses. The mutual responsibilities and contributions of people in communities were advocated by such religious writers as Paul in the New Testament, who described the community as a body with many parts, each of which contributed to the well-being of the whole (1 Corinthians 12:12–27), and Luke, who described the early Christians in Jerusalem as a community that "had all things in common" (Acts 2:44; cf. 4:32). In the thirteenth century Thomas Aquinas declared in his work *On the Governance of Rulers* that one individual "could not sufficiently provide for life, unassisted. It is, therefore, natural that man should live in company with his fellows," and added that it is necessary for individuals to live in a group to assist each other and to make different discoveries. And in the sixteenth century John Calvin wrote in his *Institutes* that Christians should not hold their gifts to themselves but must share them with others, just as they receive benefits from the community as a whole. In contemporary thought, the British scientist James Lovelock describes the earth as acting like a living organism in which life forms and chemical and physical processes interact for the benefit of the earth's ecosystem as a whole.

In all of this, there is a sense that the whole benefits from the positive integration and interrelation of its component parts. Therefore, just as a human body cannot be comprised of one single part, and just as no individual today has all the skills necessary to provide for his or her own food, clothing, shelter, medical needs, and defense, so, too, must each individual part of the social organism responsibly contribute to the well-being of society as a whole and thereby provide a benefit to society and to other individual parts of the social organism. This mutual regard and responsibility is at the core of communitarian ethics.

Some vestiges of communitarian ideals still are present in Western societies today—for example, judicial processes that remove harmful citizens from society; farm cooperatives that benefit individual farmers; and social welfare programs that assist suffering citizens in society. In the first example, the insight is evident that the community as a whole requires that its government have the authority, the power, and the support of tax monies raised from individual citizens to remove from free activity in the community those who pose a threat to the health, safety, property, work, livelihood, or even life of other members of the community. In the second example, individual citizens in a particular field of endeavor join together in a free association whose purpose is to serve as the agent for the collective benefit of all of its individual members, particularly in the area of decent prices for the fruits (and vegetables and dairy products) of their labor. In the third example, the citizens as a whole act once again through their common agent, the government, this time not to remove criminals from circulation but to help to keep in circulation those members of society who for reasons of personal, structural, or corporate inadequacies are unable to provide for their most basic needs. In all of these examples, members of the community act together to promote their interests as a community, which also coincide with the needs of individual members of the community.

The ideals of communitarian ethics could extend also to the workplace. In the United States and other countries, there is a great disparity between the salaries paid to the top executives and upper-level managers in a company and the wages paid to the workers who perform the everyday tasks without which the company could not survive. Each position is essential to the smooth running of the enterprise. In housing, for example, the most brilliant architects would never see their house designs brought to reality were it not for the workers with picks and shovels and backhoes and dump trucks doing initial excavations, and for the subsequent efforts of electricians, plumbers, carpenters, glaziers, and painters. Each person has a certain task to fulfill so that the house plans become a home, and each task is dependent upon and integrated with every other. Many years ago, there was a children's song called "We're Building a City" which taught that to build the city "we're working all together," and spoke of the different types of workers needed to construct the buildings of the city. The song emphasized the value of the work of each person and the respect due each person whose individual skills and efforts made construction of the city possible.

14 / Ethics and Development

CONTENT IN CONTEXT 1.4

Responsibility, Respect, and Remuneration

James Jones is a janitor for the Allied Auto Company. He is grateful for a steady, full-time job, but wishes the pay were higher, or that Allied at least would pick up all of his family medical insurance. "I wish I had a more important job, but there's no opening on the line." One night, as James cleaned out some metal filings left in the cogs of an assembly-line machine, he mused that if he had not removed the filings the machine would have broken down the following day, cars would not have been produced, workers and managers would have been unemployed, dealers would have lacked inventory for prospective buyers, and Allied might not be in business for long. "I'm just as important as the guys on the line!" he exclaimed. "Without my cleaning, they'd be out of a job and the customers would be out of cars!" James took extra pride then in his work, and wondered if Allied would ever pay him more than the dime over minimum wage that he received.

He looked up at the giant photograph on the wall portraying a smiling Allied CEO, and wondered if Mr. Imaboca realized how important his work as janitor was to the company as a whole. "Probably not," James said aloud with a smile, then wondered how much the boss earned per hour. "More than me, that's for sure, but he couldn't make it without me."

How often do the workers or executives in a company acknowledge their need for each participant in the productive enterprise, no matter where employees are located on the pay scale, and no matter what their particular role is in company operations? Would such acknowledgment lead to greater mutual respect? Should such acknowledgment lead to greater internal salary equity among all those who work for a company? If so, then should there be some ratio between the lowest- and highest-paying jobs: between Jones and Imaboca?

A similar attitude of respect for and appreciation of the work of each and every member in any enterprise could not fail to have a positive impact on the professional pride of every manager and worker, thereby enhancing their productivity and the overall well-being of the enterprise. A company that recognizes its need for a variety of workers with a diversity of skills, from the most basic to the most sophisticated, carefully considers the personal needs (for salary, health, safety, and rest) and professional needs (for necessary tools and workplace structure) of each contributor to corporate success, and strives to meet those needs to the greatest extent possible within the constraints imposed by company finances and productivity requirements. When each member of the enterprise is accorded respect because each task, no matter how simple to perform, is needed for the enterprise to function well, then the workplace will be characterized by collegiality, congeniality, and consequent enhanced productivity.

THE ETHICS OF TRANSFORMATION

The study of ethics can be a fascinating intellectual exercise. Exploring the intricacies of opposing positions and weighing the subtleties that differentiate complementary positions can engage the intellect in complex reflection. In such reflection, a person might try, in order to appreciate the way in which the various ethical approaches differ, to consider how these intellectual formulations relate to the material world—in other words, to analyze and judge the suitability and value of ethical positions in terms of their real-life applicability.

The ethical approach presented in these pages is to strive through the ideas and involvement of ethical practitioners to transform human society and, in so doing, to have a positive effect not only on individual and social human beings but also on the planetary context in which they live and interact. This approach is called the *ethics of transformation*. The idea inherent in the ethics of transformation is that people should be enabled, within their own concrete historical circumstances, to act with some sense of principled conduct when seeking to change conditions harmful to themselves, to others in their era, to generations yet to come, or to the earth on which all must live.

CONTENT IN CONTEXT 1.5

Transforming Transylvinya

The small nation of Transylvinya is suffering from the economic hardship that is the legacy of a four-decade political dictatorship. The Sleezoza ruling family had controlled manufacture and trade, owned most of the arable land, built up a massive fortune, and then looted the national treasury just before fleeing the country to escape retribution from the enraged and energized populace that seized power through a popular revolution.

As new leaders emerge from among the victorious rebels, they seek to determine what domestic and foreign resources might be available to stabilize their country and meet the basic needs of their people. Overtures to wealthier nations bring some promises of aid, but with a qualifier: Transylvinya must put in place an austerity program that will guarantee foreign investors a good return on their investment. The Transylvinyan government will be expected to invest in infrastructural support for business rather than in social programs for the homeless and hungry victims of the dictatorship or for the projected unemployed of the new era.

The new leaders are dismayed: people desperately need assistance in the present while waiting for the development of their new political and economic democracy, and might need transitional help in the future as that democracy unfolds. Clinics and schools are sorely needed, as well as housing and a stable food supply, to help the people realize their dream of a new and better life, the dream that led them to overthrow the Sleezoza family.

How might a nation, or an impoverished segment of the population within a prosperous nation, improve its economic situation so that all members benefit from the wealth produced by their work? Can it be done without outside aid? If not, then what might be fair terms for that aid?

The ethics of transformation is:

- Ethics for community: It is dedicated to fostering community orientation, community concern, and community well-being in human societies, while respecting the dignity, rights, and needs of groups and individual members in the human and earth communities;
- Ethics of commitment: It is dedicated to promoting community and planetary well-being through analysis of structures, policies, and practices that oppress people and harm creation, and through suggestions concerning responsible alternative modes of human conduct;
- Ethics from involvement: It is derived from the interaction of social consciousness and social activity in the context of work for the well-being of the human and earth communities. It emerges from reflection on the possibilities for eliminating injustice, promoting equitable relationships among people, and conserving the well-being of the earth and its life forms that provide the setting for human activities and deserve respect in their own right; and
- Ethics in solidarity: It is developed from the context of those most in need in the time and place of its incorporation in reflection and its embodiment in action, including politically and economically oppressed peoples, other threatened life forms, and the earth itself, whose life, integrity, and mode of existence are imperiled.

The process involved in the ethics of transformation has five steps:

1. Analysis of the social context from the perspective of community needs
2. Application of principles appropriate to this situation
3. Adjudication between competing principles
4. Assessment of respective consequences of equally valued principles
5. Action for transformation

Analysis of the Context

The individual, organization, or community that is exploring ways in which creation might be respected while human needs are being met equitably must first analyze the social context and social constructs to determine whether the operative values and practices are promoting the well-being of the general populace and of the earth. The analysis will focus on needs rather than wants to determine where needs are not being met, the reasons they are not being met, the types of natural and human resources available to meet them, and what must be done in the political and economic arenas to ensure that they will be met. The analysis will also evaluate actual and potential environmental impacts of meeting needs; explore alternative resources, products, and technologies to meet needs; and examine ways in which byproducts resulting from efforts to meet needs might safely be used or disposed of. This is to avoid creating a harmful situation in which to meet one need, another need is set aside.

Application of Principles

After assessing the social context, the individual or community determines which values should be implemented. Since these values are expressed in principles, the reflective person or group determines which principles apply in this context at this moment. Sometimes complementary principles seem appropriate, and efforts to improve the current context may be easy to make. At other times, contradictory principles vie for consideration, and people must choose the principle that seems to be most appropriate.

Adjudication between Principles

When principles are in conflict, their relative value must be weighed to determine if one principle has a higher priority than another: for example, for some people, nonviolence is a greater good than self-defense, while for others self-defense is more important. If one principle is given greater weight, then it becomes operative at this moment. If both principles have equal weight, or if competing groups advocate contradictory principles, then some effort must be made to overcome the impasse. The projected outcomes of acting in accord with each of the respective principles must then be assessed.

Assessment of Projected Consequences

If principles or people's prioritization of principles is in conflict, the consequences of each proposed course of conduct must be weighed and a judgment made as to which outcome is preferred, again with community needs and planetary well-being kept in mind. In a situation of gross injustice, as in a dictatorship oppressing people politically and economically, for example, nonviolent resistance and its potential for effecting change might be weighed against violent revolution and its potential for effecting change. In both cases the social and environmental consequences of each course of action should be considered. For the person who believes that nonviolent resistance is the only possible response to injustice, it is not a matter of choosing between tactics for change, but a matter of remaining steadfast in holding to the principle that love of neighbor precludes recourse to revolution, since revolution would cause injury to and deaths of neighbors. By contrast, the person who believes that preventing further health-related deaths and deaths by torture is a necessary good would see revolution as a less harmful alternative and would reject pacifist claims that revolutionary violence is no more acceptable than state-sponsored violence. For the person who believes that violence must regrettably be used as a last resort to prevent worse violence, the principle of love of neighbor is expressed in actions to ensure that all neighbors are freed from oppression; harm must be inflicted on the oppressive neighbor to put a stop to his or her harmful actions toward others in society. People representing both positions must make a realistic effort to project the consequences of their respective courses of action in each of two possible situations that follow: (1) the failure of their efforts to topple the dictatorship (which could bring further oppression by way of increased state violence against not only the advocates for change but against other citizens as well); or (2) the success of their efforts (which would mean that they now have the responsibility to work with the populace to create a just society characterized by political and economic democracy). People who wish to change unjust situations sometimes do not reflect sufficiently on consequences, not only the negative ones if they are unsuccessful, but also the positive ones if they succeed. Sometimes it is easier to denounce wrongs than to affirmatively announce workable solutions for correcting those wrongs.

Action for Transformation

The end sought by transformational ethics is the betterment of society within a context of responsibility for the natural environment in which every society exists. The previous steps—as many as are necessary—lead to action, based on ethical reflection, to transform the earth. This transformation will usually be done incrementally, as people realistically evaluate current conditions and take the concrete steps necessary to realize the better future that they envision in the present. At each step, people must reexamine their goals in the light of the new situation that results from action just taken. As realistic goals are set for each new step, and as the goals are achieved, social progress ensues. The cumulative actions taken, when they are based on an objective contextual analysis of needs and communitarian ethical principles, will lead to a renewed earth and restored relationships among peoples of the earth.

DEVELOPMENT

The concept of *development* might be seen from different perspectives. In a business context, development is that part of the innovation process in which a creative idea is concretized into a marketable product. In a societal context, development describes the socioeconomic/sociopolitical process by which a local, regional, or national entity utilizes its natural and human resources to a greater extent than before in order to improve its economic base, its employment possibilities and actualities, and its political stability, all in relation to other social entities and, ideally, with regard for environmental responsibilities.

In terms of societal contexts, historians, economists, political scientists, and politicians speak of "developed" and "underdeveloped" nations. In the past, it was proposed that there are stages of development, that is, that "developed" nations have progressed more than "underdeveloped" nations because of greater ingenuity, business skills, political progress, economic resources, and so forth; poor nations will progress to a "developed" stage when they acquire the political and economic skills and practices needed. More recently, analysts have come to realize that rather than stages there are subordinations of development—that is, that some nations' development is subordinated to the development of more powerful nations. The latter use their greater military and economic resources to wrest from weaker nations the goods they need at prices favorable to the dominant nations, and the weaker nations are left in basically colonial relationships with the stronger nations. The weaker nations provide basic resources at low prices and then must buy manufactured goods at much higher prices, thus being relegated to ever greater financial dependence upon, and ever increasing debt to, the stronger nations.

The situation of an *un*developed nation might be similar to that of an underdeveloped nation in that industrialization, financial resources, and natural resource extraction are lower than in developed nations; but such a nation is in a different relationship to the developed nations in that it has not (yet) come under the influence or control of those more powerful nations.

The relations developed–*under*developed and developed–*un*developed characteristic of economic interaction are not limited to international contexts. Within a single nation, there might be regions, communities, or ethnic groups in similar relationships. In such a country, an industrial region might dominate agricultural regions, a city might dominate rural areas, a corporation might control smaller companies, an ethnic majority or even minority might dominate other ethnic groups, the wealthy might oppress the poor, men's labor might be preferred to women's, and so on. The result is that internally that nation has one or a series of relationships that restrict the advancement of one or more of its citizen constituencies, much the same as in the international arena. In such a nation, coercive power might be employed that parallels the dominating tactics of powerful nations over weak nations: the police force effectively is militarized to act on one group's behalf, economic policies and tax laws favor one group over another, and even religion can be enlisted to show how one group should have preference over another.

In all three types of nations (developed, underdeveloped, and undeveloped), as well as internally within nations, the route taken to (further) development is dependent upon the degree of political autonomy and financial resourcefulness of the nation, as well as upon the availability of resources within the nation's power of control and individual nations' understandings of and practical applications of political and economic democracy. Further, some balance must be

CONTENT IN CONTEXT 1.6

Development: To Be or Not to Be?

The United Entities of Northalia, a superpower of vast economic wealth, sends emissaries to Centralia, a small nation on the American continent. The UEN government has heard rumors of the discovery of vast petroleum reserves beneath an inland Centralia lake. The UEN delegates inform the Centralian government that the UEN would like to be a partner in Centralia's economic development, and offer a loan of twenty billion dollars to help with petroleum exploration and extraction. The UEN expects, of course, that UEN corporations would be granted contracts for these industrial operations. "After all," argues the chief delegate, "we are neighbors, and what impacts one nation impacts others around it; you're practically in our backyard." The Centralia foreign minister replies that her country is a sovereign state, and that even though UEN is in Centralia's backyard, Centralia would rather bid out any development contracts, possibly with the stipulation that the winning corporation secure its own funds for development with the promise of fair returns on investment for its financial partners. In addition, Centralia is concerned about the environmental impacts of oil industry operations on or near the lake and wants a guarantee in the form of a bond that the operations will be clean and that any environmental damage will be cleaned up by the offending company. The UEN delegation leaves angered, threatening economic reprisals and political destabilization if Centralia does not accept its terms for development.

Whose development is sought by each party in the diplomatic conversation? Are the requirements set forth by Centralia fair, and are they realistic? How free should Centralia be to pursue its own course and pace of development? How might it avoid political and economic confrontation with the UEN while still maintaining its integrity and independence?

achieved in international and national contexts to meet the needs of distinct citizen constituencies.

Important questions to be discussed prior to and during development include:

- What type of development should occur?
- What will be the pace of development?
- Who will control of the type and pace of development?
- Who will benefit?
- How will resources be removed?
- What will be the social and environmental impacts of resource utilization, of infrastructural alteration, and of product manufacture, distribution, and disposal?

In business and social contexts, there are two forms of development: internalized development and externalized development. In the business context, internalized development describes the innovation process in its stages from conception of the innovative idea through production. Externalized development describes the innovation process in its stages from marketing and distribution through final product disposition as a recycled component of another product or as waste material. In internalized development, a product results from an internal intent, for example, to make an idea a concrete reality and to make a profit in its distribution; and a production process has an external impact, for example, it provides a needed good and the jobs that produce that good. The internal intent represents the company's philosophy and means of establishing itself. The external impact relates the company to society: from the business perspective, a production facility is sited in and relates to the broader social situation. From the social entity's perspective, the innovation process of one particular business—the company's development—is just one aspect of socioeconomic progress for the community as a whole. As just one facet of the entity's development it must be integrated with the innovation processes of other companies, within the parameters established by the social entity for business development.

In the social context, too, both types of development might be present. Internalized development refers to the social entity's enhancement, and externalized development describes how that enhancement interrelates with enhancements or stagnation in other social entities.

The business type of internal development will be described as part of the process of innovation; the business type of external development will be described as part of the process of implementation; and the societal type of development, which includes the interrelation between company and community from the perspective of a social entity, will be discussed as a process of integration. The first two types of development describe how a company inserts itself into society; the third type of development describes how a society incorporates a company into its social fabric in relation to its citizenry, other companies, internal and external economic and political needs, and environmental responsibilities. Essentially, innovation and implementation are private-sector activities; integration is a public-sector activity.

Since all types of development in some way or other will involve the diminution (temporary or permanent) of natural resources, and will have social impacts, business leaders and government officials, as well as workers and citizens, must consider several important questions with regard to particular facets of development:

- What need would be met by this particular product?
- Can this need be met only by this product?
- What resources will be needed to develop this product?
- Are there alternative resources that might be used?
- Is this the only production process capable of producing this product?
- How will the production process affect the natural environment and the social environment?
- What byproducts will result from production?
- What will be the final disposition of the product and of manufacturing process byproducts?

Innovation

The concept of innovation has been understood in a variety of ways. A creative person might have an innovative idea, imagining a new product or a new process; that idea might lead to innovative production, a new method of manufacturing; as production continues, an employee might conceive an innovative application of the new product, or of machinery used in production; completion of production might catalyze salespeople to devise innovative marketing to ensure the best distribution of the product. There can be an innovative theory about scientific matters; an innovative approach to putting that theory into practice; an innovative development to facilitate manufacturing; or an innovative service to tie the new product to customer needs. An inventor, for example, could devise a new computer chip; work on its manufacture; discover that more common, less expensive materials can be used to make the chip than was thought possible, and incorporate them into production; improvise shortcuts in the manufacturing process; inform customers of expanded data services conceived and created for the chip after it has been installed; and even suggest means of recycling the chip once it has become obsolete or inoperable (currently, some corporations even include plans for the eventual disposal of products, product parts, or product remnants when their utility is over).

In relation to business and technology, innovation is used in a composite sense: an "innovation" is the final product (such as the computer chip) whose realization has included the entirety of the creative process, from the conception of the creative idea through its concretization in the marketed product and to its final disposition, in whole or in part, as a recycled resource.

A business needs innovation in order to prosper or to survive. Innovation might be in the form of a new product, better quality in an existing product, an alternative method of producing that product, alternative product resources (different materials or new sources of current materials), recycling of products, and so forth. Scientific research, technological development, market information, and environmental exigencies guide business innovation. Ethics informs businesspeople, scientists, and technologists about human values and about principles embodying those values at each stage of the innovation process.

In industrial innovation, where scientific imagination and technological creativity aid in product development, ethical considerations might serve not only to influence internal and external business practices in an ordinary guidance role but also in an extraordinary catalytic role: consideration of broader social questions raised by so-

CONTENT IN CONTEXT 1.7

Integrity and Integration

Harry Joneson has worked for ten years as a city planner for the economically depressed rural community of Pine Cone, nestled at the foot of Paradise Mountain. Newly planted trees on the mountain struggle to cover the clear-cut scars that are the legacy of the Timber Resources Company, a lumber firm that operated for twenty years until trees, sawmills, wildlife, hunting outfitters, and backpackers had all disappeared from the area. One morning, Harry receives a call from Jerry Jetset of Excelsior Manufacturing inquiring about the available workforce in Pine Cone, city zoning requirements, and city policies on tax incentives and environmental waivers. Harry is enthusiastic about the possibility of a new company providing needed jobs in Pine Cone, but is a bit uneasy about tax incentives that might leave the town with little to show for its investment in streets and schools. He is also concerned about environmental waivers, since the town, having finally rebounded from the damage done by Timber Resources, now has a recreation focus, with attractive parks, motels, and tourist shops. Jetset assures Harry that he need not worry, that Excelsior has had an impeccable environmental record during its plant's ten years of operation in New York State. When Harry inquires about the reasons for Excelsior's relocation to the northwest, Jerry states that he will send details in the mail, and call a week after they are sent to set up a meeting with city officials. Harry uneasily remembers TRC as he hangs up the phone.

What questions should Harry have about Excelsior's data and proposal? What should he tell the mayor (a building contractor) and the two city council members (one owns an all-season lodge, and the other owns a sawmill converted to a cabinet shop)? If you were unemployed in Pine Cone, how much local scenery and community pride would you give up in order to get a job? If you were employed, could you explain your resistance to Excelsior to your jobless bowling buddy?

cial ethical reflection could stimulate business leaders and their scientific co-workers to view unmet societal needs as opportunities for business investment for profit and/or for promoting community well-being.

Either people or profit (or both) might motivate serious business consideration of the ideas raised by social ethics. The business and scientific communities might be moved by compassion for the needy other, or they might be motivated by pecuniary considerations. In either case (or where both are factors), successful innovation could play a significant role in eradicating social ills afflicting society.

Implementation

The process of implementation describes the manner in which innovation becomes (*a*) incorporated in a production process and (*b*) inserted into a social setting. In implementation, as in innovation, ethical considerations might be taken into account by business as it weighs its responsibilities not only toward its owners and employees but toward the general public as well: the people in the plant vicinity (the local community), the recipients of the product (the customers or users), and other people in distant places (the global community) whose lives might be positively or negatively impacted by the production, use, or disposal of the product.

There are three basic types of implementation:

- the concretization of a creative concept;
- the consociation of company and community for the realization and localization of an innovation; and
- consociation for concretization.

In the concretization type of implementation, human, technical, and natural resources—people, machines, and raw materials—are brought together to produce the prototype product and then the marketable product; the creative idea is transformed into a concrete reality.

When consociation occurs, the company which has concretized an idea works out a relationship with a community for the manufacture of the product. The consociation enables this particular product to be real-ized (made real) in this particular location. Consociation in this case involves the expansion of a company: it already has a manufacturing facility in this community or another one and is negotiating additional space for production.

Sometimes, consociation for concretization occurs. This is the case when a company has not yet initiated concretization of its innovative idea: it is exploring possible locations for that concretization or, having found what appears to be the most suitable location, it is exploring and negotiating with the community mutual expectations, benefits, and responsibilities.

Integration

The process by which a social entity (community, state, nation) incorporates the development processes of distinct companies into its own development plans and practices is called integration. In this aspect of development, the social entity determines the extent to which proposals for technological innovation and corporate insertion into a community are consonant with its own understanding of what it is, and with its own vision of what it might become. The community must evaluate what impact a company's activities will have on its present and future social and environmental settings.

The objectives of integration should be sustainable

- economic development,
- employment provision, and
- environmental protection.

In the integration process, local, regional, or national communities accept alterations to the status quo that are anticipated to be mutually beneficial to company and community. While the company seeks a site for the innovation process, the community seeks to enhance its economic base, particularly through the establishment of sustainable manufacturing enterprises and the provision of long-term employment in the higher-paying jobs available in technology-oriented operations. These types of enterprises and employment not only enhance the community's tax base and thereby help government to meet community needs, but they also stimulate the circulation and recirculation of capital resources throughout the community, in terms of peripheral or complementary enterprises, and in terms of the money spent by the company and its employees in the community. At the same time, the community must be conscious of the short- and long-term social and environmental impacts of a company's innovation and implementation. Integration is the process by which societies foster their own development. It might occur at the initiative of a corporation, which expresses an interest in locating its plant in a specific geographical area, or at the initiative of a community, which extends an invitation to a business (sometimes with appropriate economic incentives attached) to consider locating within its area.

Integration, as the social entity's process of development, requires that corporate plans are adapted to or related to the particular legal stipulations and economic needs of the society concerned. Although innovation and implementation are, to a great extent, expressions of corporate autonomy, integration requires that the corporation find means of expressing autonomy within the parameters of the political entity of which it becomes a part. In the best of circumstances, the business process of development and the social process of development will coincide, either initially or over time.

When a company proposes innovation and implementation to a community, the community representatives (elected or appointed government officials and concerned citizens) must ask several key questions:

- What will be the social and environmental impacts of this company's innovation and implementation?
- Is this company's assessment of its impacts accurate, and does it internalize all costs of the product and the production process, including resource consumption for the product and its production, as well as social, health, safety, and environmental costs?
- How will this company's innovation and implementation relate to that of other companies now in operation or competing for consideration for operation?
- What benefits will accrue to this community if it integrates the innovation and implementation proposals of this company?

Attention to ethical considerations is essential for companies and communities engaged in negotiations for technological development. When ethical values are absent, segments of the public and private sectors seek personal gain rather than social well-being, thereby imperiling the community's social and environmental well-being. When ethical values permeate negotiations, economic viability and environmental sustainability are more likely to be achieved as companies and communities work for mutual benefit, with concern for the environment and in the interests of present and future generations.

SYSTEM AND PRINCIPLES

In the chapters that follow, ethical principles offer guidelines for consideration of a variety of issues in the area of ethics and technology. These principles derive from a system of normative ethics with a communitarian orientation and a focus on responsibility. A contrasting approach would be a system of contextual ethics with a libertarian orientation and a focus on rights. Still other approaches are possible, such as a normative system with a focus on individualism and rights. The dominant approach in industrialized Western or Western-influenced societies usually has been individualism- and rights-oriented; the dominant approach in preindustrialized Western societies and in nonindustrialized societies today has been community- and responsibilities-oriented.

Reflection on the ethical principles proposed might, as noted earlier, reveal points or approaches at variance with some people's current ways of thinking. The extent of their potential to reflect objectively in a new way might well be related to their absorption by the particular culture or mindset they bring to this text. If their prior focus has been on individualism, the communitarian approach might seem too strong; in contrast, if their prior focus has been on collectivism, individual rights statements might appear too strong. In both cases, the communitarian perspective challenges people to examine their values and presuppositions. They might reach the same conclusions as before, but they should at least appreciate, if not appropriate, some alternatives to their position, and be able more readily to understand and negotiate with others. If respect for others' positions and openness to negotiating with them results from reflection on the ethical method and principles proposed here, possibilities are enhanced for people as individuals and as communities, as entrepreneurs or as community residents, to work together toward relationships that promote individual, corporate, community, national, international, and environmental betterment.

CHAPTER TWO

Individual and Society

One of the pressing concerns of ethics is the relationship between the individual and society. In any society, individual members and groups have competing claims on government services, manufactured products and nature's goods, which might be viewed as nature's "resources" when they are specifically designated or appropriated for human use. Conflicts arising over human products or earth resources may need to be resolved by government agencies. Efforts at conflict resolution rest on these key questions:

- What are the rights of the individual in society?
- What are the responsibilities of the individual toward society?
- What are the rights of society as a whole?
- What are the responsibilities of society toward individuals and groups within it?

Societies such as the newly independent states of the former Soviet Union and the nations of eastern Europe, which in the past have stressed state ownership, centralized planning, and collective action, need values and practices that foster private ownership (individual and cooperative), individual initiative, and personal productivity to complement their community orientation. Societies such as the United States and the member states of the European Economic Community, which have stressed private ownership, free enterprise, and individual action, need a broader community orientation and the exploration of cooperation and cooperatives to complement individualism and competition.

Another question that arises in the area of the ethical conduct of a society is, Should society as a whole be held to the same standard of moral conduct expected of its individual members? Some people hold that high standards should be proposed for individuals, but society must sometimes "bend the rules" in the interests of all of its members: there should be a moral person, but there could be an immoral society. Other people believe that society as a whole, as exemplified in its political and economic systems, must adhere to the same standards expected of individuals: systemic evil must be eliminated through social revision, social revolution, or even international intervention, just as individual evil must be constrained by moral suasion, law, judicial process, and social pressure.

MORAL GOODS AND MATERIAL GOODS

Two basic types of goods available for individuals and societies are moral goods and material goods.

The moral goods include such values and attitudes as love; compassion; respect for self, other life forms, and the cosmos; dignity; responsibility; personal and social liberty; justice; peace; and equality of persons. These moral goods are the foundation for the acquisition and distribution of material goods.

The material goods available within a society (or among the totality of societies) are those resources that are available, first, to satisfy human needs and second, to satisfy human wants. An individual might acquire, through personal abilities and labors or through inheritance, much more of the available goods than necessary, while other individuals lack necessities. The social fabric requires, compassion requests, and justice demands that those with an overabundance of goods reach out to those in need. When this does not occur, either extreme social conflict (and perhaps revolution) might result as the poor seek at least minimal goods; or government might force individual goods to become social goods through state intervention of one type or another (taxation, eminent domain, expropriation and redistribution, or social programs). The earth, and nations and communities of the earth, have limited goods available in their natural or manufactured state. These goods are initially social goods, and become individualized through productive processes and legal practices; they become social again in times of pressing need.

There are several ways in which those with abundant goods might extend assistance to those in need. One way is through voluntary, direct charitable aid with money, goods, labor, or service. A second way is through government agencies (funded by taxes) or voluntary organizations (funded by donations) established to assist the poor. A third way is through public or private provision of employment opportunities, housing, education, or other types of aid whereby the needy might acquire the skills necessary for self-support.

Some people offer the following analogy when discussing the possibility of meeting human economic needs with financial assistance (such as welfare) or material goods (such as holiday food baskets): Give a hungry person a fish and they will be hungry again tomorrow; give them a fishing pole (or teach them how to fish), and they'll be able to take care of their needs by themselves.

Unfortunately, it is not so simple to alleviate poverty and hunger. Questions about the availability and use of resources, the imparting of skills, and human and civil rights enter into the social arrangements in any community. The analogical solution must provide responses to such concerns as: Who makes the pole, and what resources are used in its manufacture? How will it be replaced when broken? Who will continue to teach people how to fish? How will others acquire a pole? How will bait be obtained? What means of transportation to the lake is and will be available? What access to the lake exists and will exist? Will the fisher have the right to sell or exchange the fish caught in order to acquire other goods, to consume the fish, or to distribute the fish for family consumption? What is and will be the stock of fish in the lake: how long will it be maintained, and by whom? What will be done during fish shortages? How would pollution of the lake be stopped and cleared? How will people eat in the meantime, until they learn to fish and have access to a healthy lake?

In our complex societies, the extended analogy might rather be: If you do not give a hungry person a fish, they will starve today while waiting to learn how to fish. Give a hungry person a fish and teach them how to fish and they will survive today and have the potential to survive tomorrow. Give a hungry person a fish to survive today and, so that they might survive tomorrow, also give them a pole and the means to acquire bait, and teach them how to fish, and allow them transportation, access to a clean lake, rights to fish responsibly and to keep their catch for food or exchange, and the result will be a productive, self-sustaining, integrated, contributing member of society.

CONTENT IN CONTEXT 2.1

Control of Life

Champagne flowed at the MedCon Corporation Christmas party. Two field reps, Rosenkra and Guilders, had returned from the Brazilian rain forest with a contract signed by the elders of the Xará indigenous people giving MedCon exclusive rights to harvest the taxale plant, which grows nowhere in the world except in their territory. MedCon scientists had determined that the taxale plant, whose root native peoples chew daily, prevents bronchial infection and arrests cancer. MedCon patented the plant and the chemical compounds unique to it, and marketed the new medicine derived from it.

As harvesting continued, the Xará people saw that their herb was rapidly disappearing and asked MedCon to decrease the rate of harvest. MedCon pointed out that their contract did not specify a pace of harvest. The elders countered that they had not expected MedCon to take so many plants so quickly, with no time left for regeneration, and leave so few for the people who had discovered this medicinal herb. "A deal is a deal," replied MedCon vice president John Coldcoeur, who hinted that if the Xará resisted he would have government troops remove them from their area. Meanwhile, back in the United States, MedCon sold its new medicine for $50 per daily pill, causing cancer patients to demand government intervention to give them access to the lifesaving drug. MedCon also set up greenhouses near its California headquarters and began to grow taxale on U.S. soil.

What should be the extent of access to taxale for the Xará, for MedCon, and for the general public? Should a company be allowed to patent a life form that it had no role in developing? Does MedCon have the right to charge whatever it wants for a lifesaving drug? What if taxale were essential food, and it was being exported while the Xará were starving?

Give a community, a nation, and a world those same basic needs, rights, and opportunities, and the result would be that hunger, unemployment, poverty, pollution, and wars would be diminished and perhaps even disappear from the earth.

The exercise of moral goods sometimes is tied to the acquisition of material goods. The practical exercise of compassion requires possession of material goods to help the needy other; and having a sense of dignity often requires the satisfaction of at least the basic physical needs, which requires material goods. Employment requires material goods upon which human creativity is exercised, and other material goods as payment for work rendered. The hungry person, then, today needs the fish for life and for the energy to work, and over time needs sufficient education and material goods and enforced legal rights to have work and fish and dignity tomorrow. One of the problems communities face is how to ensure the "fish rights" of all of their citizens. When some citizens who benefit from the status quo control the resources of the sea and own the beaches bordering upon the sea, it is difficult, if not impossible, for other citizens to gather the fish they need. Giving a hungry person a fishing pole—carpenter's tools, for example—is only the first step. The second step would be to make sure that a stocked lake is available and accessible—a forest, for example, whose trees are not owned exclusively by a timber company that has earmarked them for a clearcutting logging operation.

INDIVIDUAL AND SOCIAL GOODS

The moral and material goods available to society might be distributed as individual goods or as social goods, or as both. Individual goods are those moral or material goods appropriated and used by a particular person to meet primarily their needs and secondarily their wants. The benefits of the goods then accrue to the individual citizen, and sometimes to society as a whole when the individual's use of these goods benefits others. Social goods are those moral or material goods appropriated and used by society as a whole to meet primarily societal needs and secondarily societal wants. In theory, as society uses material and moral goods each member of society benefits, at least indirectly.

Ideally, when an individual is well informed and well intentioned, and is pursuing individual goods that will enable that person to develop their abilities to the fullest and use material goods wisely, the individual's pursuit of individual goods will lead to an overall improvement in social goods as well. It is in society's best interests to enable its individual citizens, to the greatest extent possible, to pursue their individual good. A community of dedicated people reaching their highest possible level of personal achievement will, in the aggregate, reach the highest possible level of social achievement. The skills of each will complement the skills of all; mutuality will be a societal characteristic; and recognition of interdependence will foster respect for others and their work, responsibility for one's own work, and realization of how individuals' talents and work function to meet both individual and social needs. In such a situation, the well-being of the community as an aggregate of individuals and the well-being of each of the community's distinct individuals can become congruent.

INDIVIDUAL AND SOCIAL NEEDS

Individuals and societies have needs which must be met in order for them to survive. For the individual, these needs include food, clothing, shelter, energy, health care, environmental sustainability, formal or informal education, social interaction, spiritual nourishment, and rest. For society, these needs include order, justice, internal and external security, equitable internal and external political and economic structures and relations, and sufficient land and resources to meet individual needs.

A proper balance must be sought between individual and social needs. Sometimes, society as a whole might make demands upon its individual members for its own interests: individual needs become secondary to community needs. In such cases, the individual still might benefit, because she or he is also a member of society.

INDIVIDUAL AND SOCIAL RESPONSIBILITIES

The individual within society owes to that society allegiance to just government; active citizenship; and responsible work appropriate to the individual's abilities. Society owes to its individual and collective citizens just government; fair taxation to meet social needs; social stability; and reasonable protection from physical injury or property damage by other citizens of society (internal security) or by citizens of other societies (external security).

CONTENT IN CONTEXT 2.2

Competing Needs

The city council of Sunny Valley is considering a petition by a local developer to construct a housing subdivision in which each home would have its own swimming pool. Sunny Valley, like other users of the nearby Sun River, draws down water from the river for residential, commercial, agricultural, and industrial purposes. The subdivision would be adjacent to irrigated agricultural land and would require diversion of some water from the Sun River system from agricultural use to pool filling and system maintenance.

Farmer MacDonald objects to the construction. He contends that there are insufficient acre-feet of water in the river to meet both agricultural needs and swimming pool needs. The developer counters that there is a housing shortage, that home construction would provide needed local jobs, and that the new homes would add to the city's tax base. Farmer MacDonald replies that his wheat is needed for food, that it provides flour for the local bakery, and that the bakery's long-term jobs and taxes would be lost to the community if his land were not in production. He also points out that the developer could construct a neighborhood pool, even one restricted to subdivision homeowners, and use only a fraction of the water that would be required for so many individual swimming pools. "Besides," the old farmer declared, "a single pool might help all those town folks to socialize and become better neighbors."

If you were a member of the Sunny Valley city council, how would you vote on the developer's petition? What would be the major factors you would consider in your decision—for example, which needs would you see as primary, and why? What about Farmer MacDonald's observations about promoting neighborliness in neighborhoods: do suburbs isolate people from each other and promote waste of resources?

INDIVIDUAL WANTS AND SOCIAL NEEDS

As social beings, most people tend to live in communities. They interact with other people to provide the goods and services necessary for the smooth functioning of society. In their quest for material satisfaction, some individuals might want a disproportionate share of the available goods and services. When this quest endangers society's ability to meet the needs of all of its citizens, individual wants must become subordinate to social needs. Here, too, the individual still benefits: when the needs of all or most are met, social stability and social strength result. The individual benefits in another way as well: although he or she might regret or resent being denied some want in order that another's need might be satisfied, a sense of security inheres in the presence of such social policies: should this individual need something in the future, that need will be met. The tone set by a society—particularly by its leaders' ideology, rhetoric, and actions—establishes a climate of acceptance or rejection of compassion for those in need, and affirms or negates practical, material efforts to meet their needs.

INDIVIDUAL NEEDS AND SOCIAL WANTS

In some instances, a majority group controlling society might impose or seek to impose upon other members of society laws or practices that are harmful to the individual and prevent individuals (singly or collectively) from satisfying their needs. In such cases, there is a conflict with the communitarian perspective that individual needs take precedence over social wants. For example, laws promoting racism or "ethnic purity," or ensuring perpetual poverty for some citizens, or inequitably taxing some segment of society, although legal because enacted through proper technical procedure, are unethical because of the injustice they promulgate. There are also situations in which a minority pursuing its wants denies to a majority its needs. Again, in a particular country, state, or region this might be technically legal, but it is nonetheless unethical. In both circumstances—injustice imposed by a majority and injustice imposed by a minority—thoughtful citizens realize that "law" and "justice" are not equivalent terms or identical realities.

In issues of business practices or technological innovations that have the potential for community harm, citizens must be alert to the possibility that a dominant majority or dominating minority among the general population or in the presiding government might try to impose its wants over others' needs. Societal harm will result if these efforts are left unchecked.

THE VIRTUOUS LEADER AND CITIZEN

The virtuous leader is the person in a position of power and authority who pursues the social good and inspires others to pursue the social good, who does not use acquired power solely or predominantly for personal benefit. Pursuit of the social good, since leader and citizen are members of society, includes by definition pursuit of individual good. The reverse proposition is less often true: subjective pursuit of individual goods, because of individual misperception of what might be "good," does not necessarily produce social good, although it may if individuals are pursuing what objectively might be seen as really good for them. The virtuous citizen is the one who, while pursuing his or her own individual good, pursues the social good as well, or at least does not harm the commonweal.

INDIVIDUAL RIGHTS AND HUMAN RIGHTS

Within society, individuals should be guaranteed the right to meet their needs. Beyond that basic human right, described in national constitutions and/or in such international instruments as the United Nations' International Bill of Human Rights, individuals might be granted civil rights: to private property, media access, and so forth. These individual rights must be exercised with respect for others' human rights and regard for others' civil rights.

GROUP RESPONSIBILITY

The responsibility of the group toward its members individually and to itself as a whole is to protect its best interests, and the best interests of its individual members. Sometimes difficult choices must be made on behalf of the group as a whole: group interests must take precedence over individual interests.

The best interests of the group are those which most promote, first, the social good and, second, the individual good. When a small

CONTENT IN CONTEXT 2.3

The Law Is the Law

The planet of Irnia was in an uproar. On its sole land mass, a long north–south island comprising sparsely populated forests and plains areas with cities and agricultural fields, the minority Rien people were marching through the streets of the capital demanding higher wages for their labor and respect for their communal cultural practices. Their banners proclaimed, "Just wages for our work!" and "Land to the people who work it!"

The Bocu majority resented the protest. After all, did they not provide housing and meals for their workers? It was true that male and female workers had to live in separate quarters, even when they were married, while they came to the city (sometimes for a week at a time) to work in a factory, or when they went to a farm to work the crops, but that was necessary to keep them focused on their responsibility to provide food and manufactured goods. "They're getting much too belligerent," observed the Speaker of the Bocu parliament. "Besides, the law is the law: the work policies were decided at the polls just this year, and were passed by a majority of the voters." If the Riens did not like the laws, the Bocu majority declared, they should just go back to the forests and not work: it was their choice. Let them find their own food and stop depleting Bocu reserves. The Rien parliamentary members countered that they would gladly work their own lands and provide their own food, but the Bocu did not pay them enough to buy private or cooperative farms to do so.

If you were a member of the Rien people, what would be your response to the Bocu contention that you should obey a law that had been democratically passed? How could you possibly object to democracy, to rule by the people? Are you advocating anarchy? What if any individual or group could refuse to obey laws they did not like?

group (e.g., a civic organization or a labor union) promotes the social good, a larger group might be inspired to act similarly. When a larger group (e.g., a state government or a corporation) acts for the social good, the largest groups (e.g., national governments, international organizations) could be catalyzed to act in a similar way. If all peoples pursued the social good, as individuals or as groups, global harmony would result.

INDIVIDUAL RESPONSIBILITY

The responsibility of an individual member of a group is to protect the group's best interests as well as his or her own best interests and to subordinate individual interests to social interests when truly necessary to benefit the commonweal. Individual members of a group might choose to subordinate their interests, or might be coerced to do so by the legitimate leadership of the group. Sometimes competing claims must be prioritized: for example, the right to resources for life over the right to resources for affluence; wheat for bread over wheat for cake; wood for a needed energy fire over wood for a wanted aesthetic fire; petroleum for farm tractors over petroleum for stock cars. Individuals, acting from compassion for those less fortunate than themselves, might voluntarily choose to use resources or part with private goods in order to meet the pressing needs of other members of the community for basic goods such as bread, a heated home in winter, and fuel for productive work. Societies, when individual members are not so forthcoming, might use the power of taxation or even expropriation in order to relieve the plight of their most vulnerable citizens.

INDIVIDUAL DECISIONMAKING

Individuals seeking to fulfill their responsibilities to the broader human community must sometimes make serious choices regarding their particular mode of conduct vis-à-vis society. Beyond societal laws (which might or might not be just) and beyond social standards (which might or might not reflect a concern for overall community propriety or well-being), individual citizens must choose courses of action appropriate to their understanding of ethical activity.

Judgment about the ethical propriety (or the morality or immorality) of a particular act involves consideration of four factors:

- the object of the act;
- the subject of the act;
- the context of the subject and the object; and
- the consequences of the subject's choice of the object in a particular context.

The object of the act is that which the subject wills to do, the thing chosen by the subject person. The subject of the act is the person who freely chooses the object, the actor who wants to make the object his or her own. The context of the act is the social setting of the act, the particular individual and societal circumstances in which the act is willed and done, including social structures (political, economic, legal) impinging on the act and individual situations (the degree of freedom and level of informed conscience of the actor; the rational liberty of the actor). The consequences of the act are the effects on the individual making the choice of the object, on society (as a whole or in terms of particular individuals other than the actor) and, where pertinent, on the planet (where environmental impacts might be felt) of the choice of this object by this subject in this social setting.

Reflection on a particular issue will help to illustrate the interaction of the four components of an act, and facilitate judgment about the ethical propriety of any act. Consider that in most modern societies private property ownership and private property control are viewed as rights of citizens. The state seeks to uphold those rights of private property through its coercive power: the police force, the judicial system, and confinement facilities. If someone violates another's property rights by taking their property, or by diminishing or damaging it so that its value or utility is lost or decreased (by destroying it or breaking it, for example), so that the owner loses its possession or control over its disposition, then the state's police and judicial systems might be called into action to pursue the wrongdoer, bring the person to trial, and fine or imprison him or her. (Other societies, it should be noted, place a higher value on community property than on private property.) These state actions are undertaken to uphold the rights of property owners as defined by law.

As has been noted earlier, however, property laws might be legal but not just. They might have been promulgated to give an advantage to a particular segment or group in society (for example, allowing only green males to be property owners, denying property rights to green females and blue males and females). Even when the laws are just, exceptions to their rigorous enforcement are allowed depending on the circumstances. For example, a man on trial for killing another person with an automobile will receive different judgment outcomes dependent upon whether or not he was a professional murderer acting for pay, a drunk driver whose ability to stop when he saw the victim was impaired by alcohol, or an upstanding citizen who saw the person on the street but could not swerve in time because of an oncoming car. In circumstances where the law itself is unjust, or where its strict application would result in injustice, things get more complicated.

Consider, in this regard, an act of theft. The object is to have food for nourishment. The subject is a poor woman who has been unable to find employment and whose children are hungry and crying. The context is evening in a city where soup kitchens and state relief offices all are closed, but where a vegetable stand with a surly owner is open to meet the needs of people returning home late after work. The consequences of the act will be that the small family will not go hungry tonight, and possibly that the woman will be arrested if she does not escape with the food. The woman asks the owner of the stand for a little food, but because she is ragged or of a different race or just because he lacks compassion for someone in need, he chases her away with a curse. If the woman knows some biblical teachings, she might remember that Deuteronomy says she and her family can take what they need to satisfy their hunger. If she has studied medieval theology, she might recall that the teachings of Thomas Aquinas declare she has the right to quietly and secretly take from another's abundance what she needs for survival, and that in so doing she is not committing theft but taking what is hers, for need makes all things common by the natural law. And if she is familiar with the Catholic catechism issued by the Vatican in 1994, she might remember that the Catholic Church teaches, with regard to the commandment "You shall not steal," that there is no theft when in the case of necessity a person must take for his or her own use the property of others.

CONTENT IN CONTEXT 2.4

Distribution and Redistribution

Robin Hood leaped from the tree, bow at the ready, and forced Lord Buckingham's carriage to stop. "I'll take your purse of gold, your lordship," he said; "you have plenty more where that came from, but the poor folk in the forest have none and need food and medicine." Buckingham grudgingly handed over his purse, protesting that he had worked hard to earn the gold: "Those poor people should pull themselves up by their own bootstraps, not take my money." Robin retorted that the poor folks had been forced from their lands and livelihoods by Buckingham's low wages and low prices for their farm produce: "If they had good jobs with security, or fair payment for their crops, they'd pay their own way." Buckingham laughed derisively, "They do not have to work for me, if they do not want to: there are plenty of unemployed out there who would gladly take jobs at the wages I offer." Robin answered angrily: "You pit one person's hunger against another's. Just because they take a job doesn't mean that they freely accept your wages. Their poverty and desperation coerce them into doing so."

Do you agree with Robin's analysis or with Buckingham's, and why? Should government be Robin Hood to provide food or health care, or to promulgate minimum wage and benefit requirements? On an international level, should the United Nations be Robin Hood to effect a more equitable economic distribution: for example, by establishing fair trade policies or by setting resource prices? Are poor nations coerced into accepting low prices for their raw materials because of the economic and political power of richer nations? How can more well-to-do individuals and nations take responsibility for their conduct toward poorer individuals and nations? Should they have an obligation to be compassionate or even just toward others?

In addition to questioning the individual ethics of the woman in this instance, one must also question the societal ethics involved: What are the structural causes of the woman's situation? Why is she hungry and homeless and unemployed? What values are dominant in society, and promoted by which group and ideology in society, that allow the vegetable stand owner to dismiss a hungry family with a curse, and that also might have her thrown in jail and deprived of her children if she is caught?

The difference between acting ethically and observing the law is very evident in this case. If the mother observes the law, her children will suffer. If she takes the food for her children, she will be breaking the law. If she feeds her children and is caught, she can hope—if some semblance of reason and compassion are present in the people she encounters in the judicial system—that she will be forgiven her crime and aided in her need.

The interaction between individual and society, then, can be very complex. Each individual has responsibilities toward the society of which they are a member and from which they benefit; and society has responsibilities toward its individual members. When each party—individual citizens and the community of citizens—fulfills their responsibilities, the individual and society both benefit, individual integrity and well-being are promoted, and harmony and cooperation permeate society.

THE INDIVIDUAL, SOCIETY, AND ETHICS

As individuals and the communities of which they are a part negotiate their respective rights and mutual responsibilities, the ethical ideas and attitudes cited earlier can inform the perspectives and positions of particular individuals and of society as a whole. This is especially true if understandings of social ethics, communitarian ethics, and the ethics of transformation are brought into play.

Social ethics, as we have seen above, analyzes the interplay between accepted principles to guide individual and societal conduct and the real-life situations of human conduct. In other words, social ethics is concerned with guiding conduct in context. The context includes the needs, rights, and responsibilities of people toward each other and toward the earth environment in which people interact. Communitarian ethics is social ethics oriented toward the rights of the community as a whole, while yet respecting individual rights—although viewing them as secondary, in general, to community rights. Communitarian ethics integrates individual rights and responsibilities and societal rights and responsibilities within a community context. The ethics of transformation, for its part, links ethical theory and ethical activity within the context of human striving for a better future for people and planet: it is intended to help create a new human society that is community-oriented while being mindful of human needs and respectful of individual human dignity, and it aims as well to help to bring into being a renewed earth that is respected, treated with care, and used in a sustainable manner.

If these ethical orientations were to become continuously operative within human thought and concurrently embodied in human action, the causes of much intrahuman conflict and of human exploitation of the earth would be eliminated. Were this to happen, a brighter future for people and for the earth could be envisioned and realized as an expected and enduring "given" of existence rather than a hoped-for ideal that is only sporadically made real.

ETHICAL PRINCIPLES

1. Individual needs take precedence over individual and societal wants.

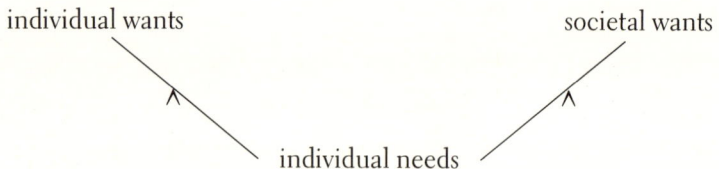

2. Societal needs take precedence over societal and individual wants.

3. Societal needs and individual needs must be balanced.

individual needs   societal needs
                             ∧

CHAPTER THREE

Employee and Employer

In many respects, the relationship between employee and employer parallels the relationship between individual and society. Similar issues emerge: the relationship of each with the other; the responsibilities of each toward the other and to the broader society.

Relationships within a corporation can inspire or impede technological innovation. Inspiration for innovation can emerge most readily in a work environment in which each member of the corporation recognizes the importance of their own role within the totality of operations; understands how best to fulfill that role; has the competence to fulfill it; is aware of the respect accorded that role by staff superiors and subordinates; and is stimulated by the company to be imaginative and creative within the parameters of their own position, with regard to other productive aspects with which they have working contact, and with respect to the company's operations as a whole.

Impediments to innovation include denigration of particular work functions; lack of proper employee training; designation of inappropriate employees for positions; disparagement of employees; and denial of appropriate incentives for responsible employees' creative reflection and action.

VIRTUOUS OWNERS, MANAGERS, AND WORKERS

The virtuous owner is an entrepreneur who is conscious of and acts according to her or his responsibilities toward co-owners, managers, workers, consumers, citizens, and the environment. The virtuous manager is the executive who is conscious of and acts according to his or her responsibilities toward owners, co-managers, workers, consumers, citizens, and the environment. The virtuous worker is the employee who is conscious of and acts according to her or his re-

CONTENT IN CONTEXT 3.1

Company Good and Common Good

Joan Bentsen is an engineer with the National Automobile Corporation. One afternoon, while observing some crash test videos, Joan realizes that the gas tank being installed on the new Firefly model is defective: the crash test dummies would not be the only ones to burn up in an accident if the car were involved in a crash in which it was hit with force by a larger vehicle. Joan is fired up about her discovery and tells the designer in chief, Terry Blazer, about it: "Lives will be jeopardized, and serious injury is a definite possibility, if we manufacture and distribute the Firefly with its weak gas tank." Her efforts prove to be unsuccessful. Blazer tells her that the company will not allow a cost overrun to redesign and remanufacture the gas tank. "Besides, Joan, only a few people might be hurt, and we can buy them off," Blazer informs her. "Retooling for a model change would not be worth the cost to NAC: it'll be less expensive to risk driver or rider injuries and pay medical, funeral or family support expenses for the few customers who get burned by their purchase."

Joan considers warning the public as a whistle-blower but is afraid of losing her job. She is worried, however, that if she were to remain silent, people might suffer injury or even lose their lives in a crash, and she knows she would feel in some way responsible since she knew beforehand that that might happen.

What should Joan do? Should her decision be different depending on whether she is a single mother with family responsibilities, or a single woman? What principles are involved and how must they be prioritized? Would Joan be responsible if she said nothing and people were hurt or killed in their Firefly? How should Joan balance company good, common good, and personal good? Should there be a support system to assist an employee who decides to be a whistle-blower?

sponsibilities toward owners, co-workers, consumers, citizens, and the environment.

COMPANY GOOD AND COMMON GOOD

The company good is that accumulation of benefits that enable, enhance, or ensure the well-being of the enterprise. Sometimes company owners and consequently their managers measure this good solely or primarily in terms of profits, especially short-term financial gains. Other benefits, such as the appreciation of the community, customer satisfaction, creativity, innovative development, employee satisfaction, plant safety, community enhancement, provision of employment, and environmental security or improvement are less often sought or not sought at all.

The common good is that accumulation of benefits that enable, enhance, or ensure the well-being of the community. The community as a whole promotes and profits from its integral well-being; its individual members are primary beneficiaries as parts of the community, but secondary beneficiaries as individuals.

Company good and common good often are complementary, with congenial relations between corporation and community. Company good and common good sometimes are contradictory, with conflictive relations between corporation and community.

Social ethical considerations weigh the common good (the good of the totality of society) over the company good (the good of a particular segment of society).

In weighing the company good and the common good, employees and employers might have to choose to support one over the other in some circumstances. An employee might have to reveal that employer processes or products threaten the common good. An employer might have to intervene when employee attitudes or actions threaten the common good.

MUTUAL — INDIVIDUAL AND CORPORATE — RESPONSIBILITIES

The individual employee, employees as a group, and employers have responsibilities toward one another and toward the enterprise of which they are a part. The fulfillment of these mutual responsibilities enhances overall operations, promotes creativity, and improves productivity. The result is a better product or service and therefore a greater likelihood of corporate financial stability and individual job security.

Individual responsibilities:
- Owners: provision of adequate capital for proper and safe operations; careful selection of managers; loyalty toward responsible employees
- Managers: responsible hiring and promotion practices; just compensation for work done; loyalty toward responsible owners
- Workers: responsible work habits; just output; care of equipment and tools; loyalty toward a responsible company

Corporate responsibilities:
- To the community: safety of operation; good products; stability
- To employees: safety of workplace; adequate equipment; security
- To shareholders: adequate return on investment

COMPANY LOYALTY

In the decades following World War II, as business prospered and expanded in new directions particularly as a consequence of creativity in technical areas, U.S. employees participating in new ventures with some measure of job security had a basic attitude of loyalty toward the enterprise in which they were engaged. In part this resulted from worker gratitude for stable work at good wages; in part it was the expected worker response stimulated by the company. There was an overall sense that, working together, labor and management could forge a competitive bond vis-à-vis other enterprises.

As time went on, however, first management then labor drifted away from the concept of company loyalty. Having found that their workers could be replaced by machines or by cheaper labor overseas, companies lost even a paternalistic concern for the job security and financial well-being of their employees. Workers and people at different levels of management came to understand, in this new employment climate, that their services might be welcomed by a competitor, perhaps with increased pay and higher job status, better job security, an enhanced work environment, or even a better locale in which to raise a family.

To a certain extent, then, "company loyalty" became an anachronism in the United States, even while it remained a practice for a time in other nations, most notably, perhaps, in Japan. (It is retained, of course, in cooperative enterprises around the world.) As a new millennium approaches and company loyalty recedes from the economic scene, business and labor both might reflect on the consequences of this new state of affairs. If workers become "transcorporate" in the same way that their corporations have become "transnational"—in both cases without a lasting tie to or loyalty for a particular place or political or social entity—what type of societies might develop, and what sort of instability (in the business world, financial markets, and national and international governance) might result? Company loyalty as a concept and as a practice seems to be an anachronism in today's highly competitive world; the question remains whether the new trend will be ultimately beneficial or ultimately harmful to business, labor, and society as a whole. The loss of loyalty means the loss of a sense of mutual responsibility, on the part of both the employer and the employee, for the enterprise itself and for the broader community in which it is situated.

Company loyalty as an ideal may be defined in two respects: loyalty of an employee or of employees to a particular enterprise; and loyalty of an enterprise to its employees. Loyalty works both ways. It must be earned by each party and should be honored by both. It implies mutual responsibilities. In the contemporary business world, loyalty is becoming less and less of a value as employers seeking greater profits at employee expense fire their higher-paid older workers, raid pension plans, and shift operations to other parts of the country or to other countries; and as employees seeking greater job satisfaction, more meaningful work, and ever better status, wages, benefits, and working conditions transfer from company to company.

In an increasingly competitive international economy, companies seem less concerned with loyalty and more consumed by the search for profits: for growth or just for survival. In such a context, companies have turned increasingly to hiring temporary workers to avoid having to pay health, retirement, and vacation benefits, or to avoid compliance with nondiscrimination or minimum wage statutes. In the more skilled professional job roles, some few employees might benefit: viewing themselves as independent contractors, living in an area where their particular skills are in demand (precluding the need periodically to move their home and family), and being able to command substantial remuneration (to pay for needs, wants, and the health and rest benefits previously provided by employers), they profit in the short term; however, age, health, and business changes eventually can cause them to be terminated as well, creating a social nightmare in the long term. For the majority of workers, the nightmare exists in the short term as well: deprived of fair wages, benefits, and job security, they suffer from instability, depression, and, increasingly, poverty. The cycle can become ever more vicious and disheartening as they compete with unemployed and underemployed others for jobs requiring lesser skills, and are coerced by employers into unjust but legal working conditions and compensatory schedules.

The effects of the corporate practices of closing operations in one community to move elsewhere in the same country or shift to another country; of firing older workers; or of hiring solely or primarily temporary workers include reduced employee creativity; psychological and financial insecurity for workers; lower employee purchasing power, with consequent retail declines, company closures, and rising unemployment; strain on public welfare services; family instability; and, ultimately, social unrest. In the short term, company profits might rise; in the long term, corporate anarchy and community hardship might result as a decreasing pool of skilled and experienced workers continually sell their labor to the current highest bidder or decide to exist on public funds rather than move around following jobs like seasonal migrant agricultural workers. When people weary of job searches give up, or are unable to find stable employment and become dependent on public largesse, the few with permanent employment often resist higher taxes for social programs, blaming the poor for their poverty. The long-term prognosis for the trend toward "disposable workers," to viewing workers not as partners and necessary helpers in the enterprise but only as expenses in the budget, is a decreasing national standard of living and increasing employee and citizen dissatisfaction.

Remedies might include restrictions on competitive foreign imports (often rendered "competitive" because of lower standards of liv-

CONTENT IN CONTEXT 3.2

Loyalty and Mutual Responsibility

The Transnational Clothing Corporation management notes in a budget review that its experienced senior workers, while producing shirts at the same rate as younger workers, receive higher salaries and benefits and thereby diminish company profits. Michael Makemor, the TCC CEO, sets up a new plant in a different state, hires only part-time workers, and closes the old plant. His brilliant move to cut costs and improve profits earns him a bonus from TCC—and also a generous salary and benefits offer at the competing International Clothiers Company. He accepts the new offer and moves on.

Joan Seamor, a cloth cutter at the new plant, reflects on Makemor's departure and wonders, "Will he be able now to put us out of business, and put me out of a job? And what about our new boss: will she shift operations to Thailand in order to obtain a cheaper labor force? I'm afraid that in business circles today our bosses tend to see us only as budget items and 'disposable workers.' At the same time, we employees are looking for ever better status and ever higher compensation. A sense of mutual responsibility for the success of the enterprise seems to be disappearing."

Should company loyalty be fostered again among employers and employees? Is loyalty a social good, a company good, and an individual good? Why or why not? What might be the implications for both Michael and Joan if mutual responsibility were to be fostered in TCC and in ICC? What might be the future prospects—for business and society—if loyalty by employers and employees were to disappear entirely?

ing and lower wages, lax health and safety laws, or government subsidies); restrictions on raw resource exports (which deprive workers of higher-paying value-added jobs); mandatory benefit provisions for temporary workers, with state assistance for smaller businesses; and cooperative enterprises (controlled by and profiting owner-workers). These remedies, which would be welcomed by most workers and rejected by many owners, would promote stable employment and national stability and security over exploitation of workers, excessive corporate profits, and the transnational profiteering that benefits a few at the expense of the vast majority.

Life in the "global marketplace" should not have to mean that working people become disposable for corporate profits; it can mean that workers in each nation work with local resources and have fair international agreements and competitive domestic markets for their manufactured goods, with consequent social well-being.

In this new global context, ethics and technology can work together to promote human welfare. Ethics can address such issues as just social contracts (employer–employee; company–community; intercorporate; international), mutual responsibilities of business and labor, stewardship of natural resources, equitable distribution of manufactured goods, and overall political democracy and economic democracy. Technology can help to achieve these social and economic goods through innovation: production of new tools and new goods, and introduction of new uses for existing tools and resources, all of which provide jobs and social wealth as value is added both to raw materials and to existing instruments of production.

Ethics and technology together can help to guide business practices and government policies to promote harmony rather than harm, stability rather than strife, and a social context in which employers and employees fulfill their responsibilities to each other while they contribute together to the well-being of the broader community.

EQUITABLE EMPLOYMENT OPPORTUNITIES

Societies that are heterogeneous include ethnic, gender, racial, religious, or other differences between citizens that either are inherent in people or originate in their personal choices or their contextual influences. Sometimes owners, managers, or personnel officers are tempted to hire, promote, or compensate workers on the basis of their own similarity to the one they are selecting for the position rather than on the basis of an objective evaluation of that person's competence for the position sought or currently held. Employee selection or rejection based on the inherent, chosen, or context-influenced characteristics of persons is discrimination.

Discriminatory practices are harmful in three major ways:

1. They hurt the individual personally and professionally.
2. They disrupt the social fabric.
3. They deprive the company of a diversity of talent and perspectives that can aid the business venture internally (through innovation, production, and marketing) and externally (through community relations and even international relations).

In some nations, such discriminatory practices are illegal, but still might occur; in other nations, they are accepted, usually for narrowly nationalistic, racial, or religious reasons.

Companies are learning to be more accepting of workers with differing characteristics as long as they fulfill their job responsibilities well. Companies as a whole and in their individual workers and managers have learned that a worker's wheelchair or sexual preferences are less important than a worker's productivity. This has come about after much soul-searching, struggle, evaluation, and legislative and legal action. Harmony and productivity in a company are promoted when people put aside their disagreement with one another's moral perspective or lifestyle, or their discomfort with another's physical appearance, and see others as co-workers with similar aspirations, abilities, and accomplishments.

JUST WAGES AND PROFIT SHARING

Competing workers with comparable skills should have equal opportunities for employment, and should have equitable levels of compensation within employment. A just wage is compensation that enables employees to meet their basic individual and family needs, including food, clothing, shelter, energy, and medical costs; that provides salary or benefit adjustments relative to workload adjustments; that is comparable to remuneration received by other employees in equivalent positions in the same enterprise; that enables them to set aside some savings for future needs and for recreational purposes; and that provides for periods of rest and recuperation.

In some businesses, the ability of employers to provide such a wage might need to be subsidized through government aid either to the employer, in the form of tax credits or health insurance assistance, or to the employee, through social welfare programs providing food, housing, medical services, and so forth. Government might also increase employers' ability to pay a just wage through restrictions on the unfair foreign competition for domestic companies embodied in some imported products. The sale or dumping of foreign-manufactured goods, coming from countries having lower labor costs or few if any environmental safeguards, or supported by government subsidies for business, hurts domestic production. (Such trade restrictions should not, however, be employed merely to atone for poor planning or shoddy manufacturing by domestic producers.) An alternative to trade restrictions, in nations believing themselves to be facing unfair international competition, would be to develop good public–private partnerships in which government revenues and research capabilities would balance government regulatory efforts, complement corporate resources, and catalyze business innovation.

Sometimes the employer and employees, when a company is just beginning or when a difficult financial situation arises, might negotiate together temporary levels of compensation, below the level of what might be considered ordinarily a "just wage." The negotiation should include provision for "just" wages and benefits when the company position is more stable, and improved compensation when the company prospers: employees who take cuts and risks and share in hardships should share also in the wealth from stable and prosperous periods.

Some businesses have implemented programs of profit sharing with their employees, an opportunity for employees to benefit from their productivity in a way complementary to their salary. Profit sharing benefits both employer and employee: since the employees' share of profits is a percentage of the total profit, they tend to work harder and more carefully, thus generating more goods at less cost. Employers receive a better financial return on their investment in the form of higher earnings; employees receive a salary bonus.

CONTENT IN CONTEXT 3.3

Just Wages

Pacific Pen Producers responds to an increased demand for its new computer mouse pen by opening a new facility in Inkberg, Tennessee. The city and its surrounding county have been experiencing high unemployment because of failed businesses and farm foreclosures, and new jobs are eagerly sought by area residents. PPP, noting the surplus of potential employees, decides to offer to job applicants only minimum wages and minimal benefits, although the company would make a good profit even if it paid its workers wages comparable to those at its Minnesota plant. PPP also considers closing that plant and shifting all of its production to labor-laden Tennessee. "It just makes good business sense for us to make more money here," explained owner Gerri Graphic to Mary Constant, a disappointed applicant. "You don't have to work for PPP if you are unhappy with what we're offering you."

Mary mused, "It seems the concept of a just wage for responsible work has eroded in some businesses like PPP. What recourse do we have as workers? If we object to low wages before we're hired, we won't get the job we seek. If we object later, we'll be called agitators and get fired. If our unionizing efforts are successful, the company might then move overseas, taking our jobs with it."

What principles and priorities might serve as bases for negotiations for a just wage today, and might be adaptable in the future? What should be the role of government if a particular business cannot or will not pay a just wage and provide corresponding benefits for its employees?

Another form of compensation that might be given to employees is a share in the stock of the corporation. Like profit sharing, this arrangement can benefit employer and employee because of enhanced worker productivity to increase stock values. Unlike profit sharing, there is an element of risk involved for both employer and employee because of stock market fluctuations: employees might sell their shares in hard times to meet needs, or in prosperous times to profit from higher prices per share; employers might make decisions that lower the value of the stock, thereby diminishing workers' earnings. One employee stock ownership program that some companies offer to offset these risks—and that both gives employees a stake in their company and promotes company stability—is to have employees retain ownership of their stocks as long as they are employed by the company, and to sell their stocks back to the company at market value when they leave or retire from the company, or when their employment is terminated by the company.

PROPORTIONAL COMPENSATION

In some enterprises a policy of proportional compensation has been established. This means that there is a ratio incorporated in the remuneration provided to the highest- and lowest-paid employee. This ratio might be as close as 3:1 or as separated as 7:1: the highest-paid employee receives as little as three times or as much as seven times the compensation of the lowest-paid employee. The underlying premise here is the recognition of the value that each employee has for the enterprise, whether chief executive officer or janitor; each needs the skills and work of the other. For example, in an industrial operation, if the janitor did not sweep the floor at night, metal filings might enter the machinery, assembly lines would close down from breakage, designers and marketers would have no reason to work, and the CEO would have no one to manage. On the other hand, the CEO's proper management guides the company to a position of stability, enabling the janitor to have a job.

The advantages of proportional compensation are that it accords meaning to each position and respect to each worker; eliminates gross disparities in compensation; lessens class divisions and jealousies based on job status or compensation; diminishes quests for a different position solely for monetary or status reasons; and promotes greater congeniality, collegiality, and community within the enterprise. The potential disadvantages of proportional compensation are that some employees, feeling that their contribution to the business merits greater compensation or higher status, cause internal dissension or seek employment externally to obtain such rewards, thus disrupting the internal operation of the enterprise.

COLLEGIALITY AND COOPERATION

Teamwork is essential for the successful operation of any enterprise. Collegiality means that work roles are assigned and fulfilled in an interactive and interdependent manner: the corporate structure is such that the distribution of work means shared responsibility. Cooperation implies that as the respective members of the enterprise work collegially according to their respective roles, they assist each other in the fulfillment of those roles.

SUBSIDIARITY

When members of an enterprise are aware of their respective and mutual responsibilities within the enterprise, their work should be so organized that each fulfills his or her responsibilities at the highest levels possible. Putting it another way, each member delegates or receives work as far down the structure as is feasible given the abilities and roles of others, so that those with greater responsibilities are not burdened with tasks that might be carried out at a lower level. The result of subsidiarity is greater productivity and greater opportunity for employees to operate at the highest possible individual level, thereby enhancing self-worth and improving job satisfaction.

COMPATIBILITY

Managers and supervisors should take care to ensure that interactive teams are internally and externally compatible, able to work well with each member of their own and interacting groups. Productivity will be increased in direct proportion to the compatibility of interacting teams and of members of individual teams.

CONGENIALITY

Employees should strive for a friendly and respectful atmosphere at work, and employers should try to provide the conditions that encourage such an atmosphere. Congeniality allows for mutual accomplishments through considerate give-and-take and overrides irritation from distinct personality types. It decreases internal competition and conflict and enables the enterprise to work as a cooperative team.

COMPETENCE

Employees should be assigned roles consonant with their competence; employees should not seek employment in or advancement into positions for which they lack the skills or interest, unless a mutually agreed-upon training program is included as part of the initial training or immersion in the new role they seek. When people thus realistically assess their interests and abilities, they will be happier in their work because their role will be commensurate with their skills; neither they nor their supervisors or peers will experience frustration over work incorrectly performed; and the enterprise as a whole will function more smoothly. Managers should assist in this realistic assessment of the potential of their subordinates through interviews with co-workers and/or aptitude tests and interest inventories.

COMMUNICATION

Effective operation of any enterprise requires effective communication among its employees, branches, divisions, and so forth. The whole enterprise suffers if this beneficial communication is impeded by (1) individual egos, incompetence, laziness, or turf concerns, (2) extreme competitiveness or jealousy and withholding of ideas or information, or (3) lack of appropriate structures or equipment.

Communication stimulates imagination and innovation. People engaged in different tasks know firsthand the difficulties encountered in fulfilling their particular role in the enterprise. When encouraged to do so by bonuses, company pride, recognition, profit sharing, and similar incentives, employees will use their knowledge of what their jobs require to seek creatively to improve their productivity. Money can be saved, safety improved, competitiveness retained, and profits increased through communication practices that encourage and reward employee creativity.

Good communication contributes to job satisfaction. When employees understand what is expected of them, they can more easily fulfill their assigned roles in the business structure. In addition,

CONTENT IN CONTEXT 3.4

Beneficial Communication

One sunny Wednesday morning in the Los Angeles plant of the Cascade Car Corporation of California, Maria wonders if the electric automobile that she is assembling could be modified to incorporate solar cells as the major power source, with a plug-in to make the bulkier batteries the backup source of electricity. The idea came to her when she was balancing her checkbook on a solar calculator that had a battery backup. She reasons that in southern California and other areas with abundant sunshine the dual-power vehicle would be less costly overall, less limited in use because less time is needed for recharge periods, and more viable and more environmentally safe. Maria approaches Joe, her foreman on the line, with her idea. "What do you think, Joe? Will the guys in design be open to some modifications? It might give us a competitive edge if we have the first solar car that's economically feasible and environmentally safe." Joe considers Maria's idea. He remembers that Cascade offers bonuses for good ideas, and this one just might fit into that category. "I'm no engineer," he replies, "but it looks like it could work. I'll think it over and get back to you."

What would be the impacts on Maria, her company, energy consumption, and the environment if these are the possible reactions to her idea: (a) Joe points out that she is not an engineer and should not get into an area she knows nothing about; (b) Joe takes her idea to the plant manager as if it were his own; (c) Joe encourages Maria to discuss her idea with the plant manager; (d) in b and c, the plant manager (1) dismisses the idea outright; or (2) passes it on to the company engineers for exploration and feedback.

How might communication be improved between Maria and the engineers? How might a company balance its quest for good ideas with its fears that some employees might offer bad ideas from outside their field of competence and waste other employees' time?

where a proper grievance procedure has been established, company members realize that conflicts can be equitably resolved. Finally, all can have a sense of contributing to the proper functioning and well-being of the enterprise if they believe that their good ideas will be incorporated into the productive process. Their sense of responsibility for the venture increases, helping them both to endure hard times and to enjoy prosperous times with their company.

COOPERATIVE ENTERPRISES

The two major economic systems today, capitalism and socialism, have different areas of focus. In capitalism the focus is on the individual good and individual freedom; the common good is secondary because it is assumed to be promoted by the sum of individual goods. In socialism, the focus is on the common good and common needs; the individual good is secondary to the interests of the whole. Each system has advantages and disadvantages in terms of satisfying human needs and wants, and of balancing social and individual goods.

The advantages of capitalism are that (1) it recognizes what seems to be a common human need: ownership of private property, with the apparent security that property brings (a home in which to live; tools for work; a farm to provide home and livelihood; one's own business; a car for mobility); and (2) it enables the citizen to make choices for goods from a wide range of options. The disadvantages of capitalism are that (1) ownership can be concentrated into very few hands as individualism in the form of selfishness is elevated to a virtue; and (2) excessive production of goods to satisfy citizen demands (some of which are wants created by advertising) creates environmental exploitation and resource waste.

The advantages of socialism are that (1) it recognizes that people can work together for the common good as a social enterprise, and have a sense of security because of state provision for their necessities (housing with controlled costs; free health care; free education; guaranteed employment with compensation sufficient for needs satisfaction; adequate public transportation); and (2) its planning process and production offer essentially needed goods and some wanted goods, limiting resource waste. The disadvantages of socialism are that (1) nationally, important production and distribution decisions are in the hands of a few people with little knowledge of local needs, problems, and resources; and (2) although some citizens are motivated by altruism or enlightened self-interest to work for the common good, others yearn for greater individualism or lack work responsibility because of guaranteed employment and the lack of financial or status incentives for work done well, thereby decreasing the quality or quantity of production of goods for society.

Some nations have a mixture of capitalism and socialism. In Canada and the United States, for example, the basic structure is capitalism with some socialist elements—for example, free education and socialized medicine in Canada, free public school education through the twelfth grade, Social Security, and other social service programs in the United States. Sweden, with a basic socialist structure, has the world's highest standard of living, and contains elements of capitalism; and the states formerly part of the Soviet Union currently are in a transition between a state-controlled economy and a mixed economy using aspects of U.S. capitalism such as increased private ownership of the means of production and economic competition with producers of the same manufactured good or agricultural products.

An economic structure that contains the best elements of both capitalism and socialism is cooperativism. The cooperative enterprise includes:

- private ownership by its members, satisfying the human desire to possess;
- a socially cohesive work unit, focused on creative cooperation;
- an equitable distribution of the financial gain ("profit") realized through work, promoting internal respect, productivity, and stability; and
- a concretized social consciousness, oriented externally toward fulfilling the needs of the broader community.

Cooperativism already exists in both capitalist and socialist countries as a small part of the broader economic structure. In the United States, for example, farmers' dairy cooperatives abound in the Midwest; in Nicaragua under the Sandinista government the "tear down the fences" program gave financial incentives to small farmers to work their still privately owned individual lands as a cooperative and divide their market revenues proportionally.

Cooperative agricultural holdings, as well as cooperative manufacturing plants, might be located in the same area or even on adjoining property, or might be widely separated. In the first instance, they might be worked more easily and profitably as a joint single enterprise; in the second, as consociated enterprises that buy resources jointly, ship goods from one to another for value-added production, and/or market their products as if they were from one enterprise and divide the profits in proportion to their respective input. Unions of cooperatives also exist, most notably the Mondragón cooperative in the Basque region of Spain. In such cooperatives, each group of similar enterprises has its own cooperative, linked in a federation with other cooperatives and having an umbrella governing board. When the members have a developed social consciousness (as in Mondragón), individual cooperatives can weather hard times through the assistance of cooperatives who are prospering because they are producing a different good; diversification enables each to support the other as demand for different products fluctuates from market changes. This social consciousness also stimulates the members to use part of their excess of revenues over expenses to support public needs through community projects and, in some cases, to limit individual incomes in the interests of the common good, social harmony, and respect for every worker's dignity and contributions, from whatever task, to the cooperative enterprise.

INDUSTRIAL COOPERATION AND ESPIONAGE

In order for society and business to benefit from the best possible innovative ideas and practices, corporate interaction is often necessary. A certain synergy can result from external business relationships such that the combined cooperative output is greater than the sum of the outputs of the individual enterprises involved. New products, new modes of production, greater environmental benefits through such practices as recycling secondary products from one operation to another, all these might result from corporate cooperation on a regional, national, or even international scale.

A potential danger from cooperative corporate efforts, which also is present in competitive corporate efforts, is that technology might be surreptitiously, unscrupulously, and illegally transferred from one company to another. The true inventors in such cases lose their legitimate right to benefit from their innovation, which might have occurred either through their own entrepreneurial skills or by assignment of aspects of the innovation to others for development with appropriate compensation to the inventors. Industrial espionage robs the fruits of imagination and production from those who labored to produce them. The theft of intellectual ideas or of manufactured goods promotes tension, mistrust, and conflict among companies and nations, and eliminates possibilities of interaction and cooperation. Harm results not only for the company suffering the loss of its property but eventually for the offending individuals and company as well. In contrast, mutual respect for the intellectual and manufactured property of others leads to cooperative production and distribution efforts beneficial to innovators, entrepreneurs, manufacturers, communities, and countries.

INTERCORPORATE AND INTERNATIONAL CONTRACTUAL OBLIGATIONS

A business consideration complementary to the threat of industrial espionage and theft is the development, understanding, and implementation of business contractual obligations between corporations and between nations. When fair contracts are mutually accepted by corporations establishing a particular business arrangement, all parties to the contract are morally and legally obligated to observe the conditions of the contract. The trust essential in business circles for the cooperation that leads to shared innovation is negated by contractual ruptures by any or all parties. Business and society both ultimately suffer when this becomes the case. Eventually, fear of losing business or of losing the legitimate benefits of innovation causes a general deterioration of cooperative efforts. The synergy that might have resulted from such efforts is not realized; competition accelerates; and losses occur.

On the international level, contractual responsibilities are no less obligatory than on the business-to-business level. In the business arena in the past, nations competed with each other for markets or for a technological edge. While this continues to a great extent today, nations—and the companies within nations—have begun more cooperative efforts; space programs represent one area of such cooperation. National governments and even nationalist entrepreneurs

CONTENT IN CONTEXT 3.5

Equitable International Agreements

John Johnson from Earth and Axzl Axzlrq from Pluto are negotiating the sale of Plutonian mercury chips for computer use. It is Axzlrq's first trip to Earth, and she asks Johnson about the customs requirements for alien cargo. Johnson, hoping to make a fast buck, deceptively informs Axzlrq that Earth customs procedures require that off-planet imports be charged a docking fee, payable to him. As Axzlrq pays him, Johnson simultaneously thinks about a New York manufacturer who could replicate the Plutonian chips and, with minor adjustments, sell them as Earth look-alike products. Johnson hopes to make millions of dollars with pirated chips. "I can even put my name on them: after all, I am an engineer, and no one would know that I didn't really invent them. I'll call them Johnson Jetchips." Unknown to Johnson, Plutonians have developed telepathic skills that help them communicate over short distances without the need for communication devices and also assist them in ensuring honesty in international and interplanetary trade negotiations. "Hmmm," responds Axzlrq.

What should Axzlrq do in her negotiations with Johnson? She realizes that if she does not pay the bribe, she will have no access to Earth markets after her long trip; if she pays it, her product will be copied and her company's and her planet's rightful royalties lost. If Johnson carries out his plan, how will that affect future efforts by Earth to secure the more advanced Plutonian technology? What types of international patent laws should be enacted that would help traders like Axzlrq as well as inventors of new commercial products to safeguard the integrity of trade negotiations, ensure fair compensation to inventors, maintain the manufactured quality of a company's products, and promote fair trade in new products between nations and planets?

realize that there are mutual benefits to be realized through such cooperation—shared knowledge benefiting all, synergy in development, cost savings for individual companies and nations as facilities and personnel are shared, and so forth. When a party to such sharing disrupts the process through contract violations, industrial espionage, or unlawful technology transfers, the benefits of integrated cooperation are lost and the individual offender in particular becomes an untrustworthy choice for inclusion in future cooperative ventures. This in turn harms not only the offending party but its nation and society as a whole.

INTERCORPORATE AND INTERNATIONAL OPENNESS AND DISCLOSURE

When competition remains the guiding principle for intercorporate relations, business owners and managers are tempted to withhold important information about their innovations, products, and operations not only from business rivals but also from potential or actual creditors and joint venture partners and inquiring potential new owners. In their search for improved business operations, joint business ventures and international business partners, these unscrupulous, uninformed, insecure, or inexperienced owners and operators of companies strive to benefit themselves or their enterprise by concealing vital information about themselves and their operations and simultaneously culling all the data they can from more forthright or honorable entrepreneurs and managers about their operations.

Lack of openness and full disclosure in these relationships can lead to further unfair business practices, facilitate flawed agreements, precipitate legal entanglements, jeopardize the security of the outside interests, and set the stage for further cutthroat business practices.

Truthfulness rather than half-truths, and disclosure rather than dissimulation, will lead to equitable business arrangements, greater intercorporate trust and cooperation, and improved business stability. The consequences will be a better business climate, improved productivity, more prosperous national and international economies, and more peaceful and productive international relations. Corporations will be able, in such an open and cooperative climate, to become collaborators as much as competitors.

REGIONAL AND INTERNATIONAL CULTURAL DIFFERENCES

As trade begins, continues, or expands among nations or between ethnic groups within a nation, differences in cultural perspectives and perceptions might be the source of misunderstandings and conflicts. Words, gestures or their interpretation, or disparate levels of significance attached to comments about business practices or during negotiations can help or hinder transactions. For some people, for example, "yes" means affirmation of and agreement with a proposal; for others it can mean "maybe," while the official or office considers the proposal in more depth; and for still others it can mean "no," stated as "yes" only in order to avoid hurting feelings, with the hope that the other party will realize from continual "yeses" or postponements of contact, with no compliance with what was thought to be confirmed, that the affirming party has no intention of fulfilling what appeared to be an agreement.

Because misunderstandings and misinterpretations can easily occur in negotiations between people of different cultures, it is important for businesspeople to study differing cultural perspectives in depth or to be assisted by trusted guides from those cultures. To do less is to invite the possibility of insulting hosts or guests at the time of interaction, of misinterpreting the meaning attached by others to statements about a proposed or signed agreement and its implications, or of having unrealistic expectations about fulfillment of the agreement in the future. The consequences will be disagreements about contractual responsibilities, possible nullification of contracts, and loss of future business opportunities for trade and cooperation not only between corporations but between countries as well, thereby inhibiting both productive business relations and good international relations.

One problem that sometimes arises with regard to cultural understandings is the use of payments to government or business officials to "open doors" or to have an "edge" in bidding or even to guarantee securing a contract. In the United States and some other nations, such payments are considered bribes and are illegal. In other countries, they might be officially illegal but overlooked; officially illegal but accepted secretly; or officially understood as a way of doing business. For nations whose officials demand bribes, and for companies whose owners or managers offer or accept bribes, the business deals made in such a way do not assure agreements in the best interests of business or nation, only that the representatives of the company with the most money available at a particular moment will receive the contract.

Sometimes bribes are made to assure overcoming competition (to preclude others' entry into a particular market, or to guarantee reception of a contract when others are bidding as well); at other times bribes are used to acquire access to markets or to assure delivery of goods. In the former case, bribes serve to undermine cooperation and true competition, and are an unjust means of assuring acceptance of an offer independently of the real value of the accepted offering or of alternative offerings. In the latter case, where payments are an accepted and/or expected part of business dealings, bribes might be an unfortunate and necessary short-term evil requisite for business until standards for intercultural and international business are agreed upon, promulgated, and implemented.

ETHICAL PRINCIPLES

1. The dignity of employers and employees must be respected.
2. Employees must labor conscientiously to fulfill the justly negotiated quality and productivity requirements of their position.
3. Employers must pay a just wage to their workers.
4. Employers must provide a safe and healthy work environment.
5. Employee hiring, advancement, and compensation should be based on competence and experience.
6. Employees must fulfill their quality and productivity responsibilities to owners and consumers.
7. When appropriate and feasible, employees should share in the ownership of the enterprise.
8. Inventors have the right to benefit from their innovation as intellectual or technical or productive property, and the right to protection from the unlawful and unethical theft of those types of property.

CHAPTER FOUR

Corporation and Community

Technology that is concretized through innovation can have a positive or a negative impact on local and distant communities. Every creative idea and application of technology emerges from a member or members of a social community and, in turn, impacts that community and others. The social impacts of technological development include creation or diminution of employment, infrastructural alterations, tax base impacts, and environmental effects. Among these social impacts, community environmental effects are usually the most neglected and the most denied.

COMPANY GOOD AND COMMON GOOD IN CONTEXT

In most contemporary social contexts, some accommodation is sought between the good of a business or technological venture and the good of the community in which it hopes to be located. Sometimes, entrepreneurs and government officials have conflictive negotiations because of apparent value conflicts (e.g., corporate profitability vs. community health), leading to government restrictions on business because of community concerns. In contrasting cases, entrepreneurs receive from overly accommodating government officials concessionary agreements that are beneficial for the company but far less so (in terms of long-term potential economic impacts and/or health implications) for the community. The best situation, of course, is one in which negotiations are marked by compromise: considerate company representatives and conscientious government officials work together to create a responsible arrangement that promotes both the company good and the common good.

Consideration of technological impacts in this regard should include frank discussion of two questions:

> Has the corporation in the past accepted development, distribution, or use of a product potentially or actually harmful to the public or to the environment?
> Has the corporation negatively impacted the environment or community health and safety through its past manufacturing processes?

If the answer to either or both questions is affirmative, the corporation should explain how it proposes to safeguard the public and the environment in its present proposal. Public awareness of product dangers, stimulated by consumer groups, government inquiries, or manufacturer recalls, can cause consumers to consider: (1) Did this company design a flawed product because of engineering incompetency, thereby endangering my (or my child's, or my family's, or my community's) health or safety? and (2) Did this company market a flawed product because of greed, thereby endangering my (or my child's, or my family's, or my community's) health or safety? In both cases, consumer well-being is at risk, and the company should be held accountable for its past shoddy design or crass profiteering and constrained from such activities now and in the future. In the long run, educated consumers might voluntarily avoid buying goods or services from the offending company, and uninformed consumers might be assisted by government intervention through the judicial system to escape victimization from the offending corporation, with the consequent loss of business by that company negating whatever financial benefits accrued from its flawed product, production, and practices.

PROFITS AND PEOPLE

A business enterprise needs to acquire, as a consequence of its work, some excess of receipts over expenditures. Whether this margin is called "profits," "net surplus," "labor-added value," or "fruit of the people's labor," it is needed to continue operations, to fund innovations, and to provide for necessary growth.

In the pursuit of profit, those directing the course of company operations must bear in mind their relationship to the people employed

CONTENT IN CONTEXT 4.1

People and Profits

Bill Watson works for the GermZap Incinerator Corporation, a company that contracts with hospitals and clinics to safely dispose of medical wastes. After work one day, as he walks from the plant to his car, he smells a slight antiseptic odor. He looks toward the smokestack of the GermZap incinerator and notes a grayish plume drifting skyward. Bill remembers that his company had informed his community, prior to locating its plant near the school on the city's edge, that the incinerator burned wastes at such a high temperature that nothing toxic would remain to endanger citizens. That statement, and the promise of jobs in a depressed area, had convinced city officials to approve the plant and grant zoning variances and property tax breaks.

As he drives home, Bill recalls two previous occasions when he had smelled the burning waste. He wonders anew if he should raise the issue at work again (last time, his supervisor had become angry and talked about lost jobs), or with friends in the community. He is worried about his health (he would have no job if he became ill), about his children in the nearby school, and about his neighbors and friends: all might be hurt if he were to say nothing. On the other hand, if he were to inform others about a possible health hazard, he might lose his job, or the community might lose the plant and its economic benefits and become economically depressed again.

Bill faces a dilemma: "If I speak up, the incinerator might shut down, and I couldn't pay my mortgage, feed my family, or have medical insurance premiums paid for at work. If I say nothing, my kids might get very sick from medical waste smoke."

How might Bill balance company good and community good, people and profits? If he speaks out, what should GermZap do? What should the city officials do if GermZap does nothing, or if it threatens to close down if the city demands improved emissions?

and to the public concerned. A consideration of human needs in both cases should lead to the realization that human well-being takes precedence over money:

$$\text{people} > \text{profits}$$

Again, profits are necessary and helpful to people. The employees of a company need that company to profit in order to meet their own needs. The people in local communities need that company to profit in order to meet local employment and infrastructural needs through jobs and taxes. The broader public might need that company to profit in order to satisfy pressing personal or social needs. The issue, then, is not profits per se: it is the manner in which they are pursued and obtained.

The principle "people before profits," as exemplified in the symbolic notation "people > profits," means that companies and communities must consider whether worker and/or citizen health, safety, or dignity is endangered or ignored by company practices. If they are, then company and community must work together and take steps to safeguard the life and health of the community.

JOB CREATION AND JOB LOSS

Companies seeking to locate in a community offer the prospect of increased employment for citizens of that community, an obvious community good, especially in more economically depressed areas. Such job creation possibilities should be realistically expressed: some companies unscrupulously promise more jobs than they can deliver in the hope of gaining greater incentive commitments (such as property tax breaks or zoning variances) from the community. Job creation is, overall, a community benefit, although communities must carefully evaluate potential strains on the community infrastructure: the need for new schools and teachers; better roads and their maintenance; increased water supply needs; increased sewage and waste disposal; and so forth. These must be balanced against the potential of the incoming enterprise to provide income to the area sufficient to more than offset new community costs.

While job creation is often a community dream, job loss is usually a community nightmare. Suddenly a community must worry about how to continue paying for that new school; which teachers to termi-

CONTENT IN CONTEXT 4.2

Community Progress?

The Transformed Toxics Corporation sends its public relations team to Big Sky City to meet with government officials and business leaders to discuss construction of a garbage- and hazardous-waste-burning facility adjacent to its Happy Homes subsidiary's cement manufacturing plant. Conrad "Con" Creet, the TTC vice-president for community affairs, promises new jobs for Big Sky City, and declares further that if the incinerator is not approved, "Employment at the cement plant will be jeopardized. The costs for Happy Homes to continue in operation to provide cement for homes and driveways would be prohibitive if we couldn't use cheap toxic wastes as fuel." Jane Morrelli, a local environmental activist, disagrees: "TTC is making more money from toxic waste disposal contracts than it is from cement production. The cement plant is your smokescreen for making big profits in your real business." Workers fearful of losing their jobs and homes, however, support TTC's efforts in arguments before the Big Sky City Commission. Jeff Nelson, a foreman in the cement plant, voiced the workers' position: "Surely TTC would not propose an incineration process that would be hazardous to the health of its employees and their families. Besides, the federal government approved the process, so it must be okay. Those environmentalists don't know what they're talking about. Anyway, it's not their jobs on the line."

What should be the response of Big Sky City residents and government officials? What analysis of the situation might they make to help their decisionmaking process, in terms of scientific testimony, decreases in waste generation, waste disposal options, alternatives to portland cement, and employment attraction? How might employment needs be met without the issue being jobs vs. environment, or community health vs. community employment?

nate because of student losses as unemployed workers move away; how to absorb sudden unemployment not only from the departing firm but also from other businesses built up to provide goods or services to employees of the departing firm; how to pay for site cleanup for health hazards left by the firm; and so forth. Job loss can prove to be more economically harmful to a community than the previous job creation was economically beneficial.

Some communities that have offered substantial incentives to companies and/or whose economy is substantially dependent on workers' salaries from particular companies have sought to get back the value of tax incentives by suing firms announcing relocation plans, or even have declared eminent domain on company assets to prevent their relocation to another site, in the latter case asserting that the venture would be run as a community corporation. With the former tactic, court cases sometimes might be won to retain jobs and save the local economy; with the latter tactic, companies might change their minds and remain in the community.

Sometimes unscrupulous entrepreneurs pit economically depressed communities (or states, or even nations) against each other in order to obtain incentives that make the business more profitable but jeopardize citizen health and the long-term financial stability of the potential locale. Such companies might then also try to relocate to a different site when the community objects to harmful impacts on citizen health, or when the time comes for the community to begin to realize some return, through taxes, on its investment in the company as represented by the tax breaks or zoning variances granted previously.

Job creation and job loss, then, are two issues with which both communities and companies must struggle. When each side looks toward promoting the overall social good, and to preventing long-term harm for either party, issues can more easily be resolved.

COMMUNITY DEVELOPMENT

The goal of community concessions to corporations is community development. Ideally, such development would not fall into the "boom or bust" category but be an economically viable, empirically sustainable, and environmentally respectful enterprise appropriate to the people and the resources of the community over the long term.

Development is economically viable when the human (individual and community), natural, financial, marketing, technical, and byproduct resources used in production and distribution can be so integrated as to enable the responsible manufacture of a material good needed or wanted by the broader public, and be capable of providing some surplus of revenues over expenditures.

Development is empirically sustainable if, based on available knowledge and experience, it appears likely that the economic viability can be extended over time without harmful impacts on the local community, the regional environment, or the consuming public. This "time test" is a crucial issue in community development plans, because otherwise the community might make infrastructural investments or offer tax incentives to assist a company, expecting to regain its financial output over time through the taxes paid by the company and by other businesses established or expanded to respond to employee needs and wants—and then lose its investment because a product is no longer demanded, or because area resources no longer can sustain production.

Development is environmentally respectful in a local area when product manufacture and utilization, and community alterations, use natural resources conservatively, with minimal impact on life and natural ecologies.

INFRASTRUCTURAL STRESS AND SUPPORT

Business can bring both stress and support to the social infrastructure.

Stress results when the impact of an innovation places a strain on the resources of the local community or region. Sometimes the local or national government has the resources through taxes to support improvements to roads, sewers, water supply systems, dump sites, airports, and so forth, thereby aiding corporate profitability through taxpayer assistance. But sometimes governments or citizens refuse to provide such support, expecting the business and its clients to pay for infrastructural changes the venture needs and by which it would profit. Where both company and community would benefit, some equitable arrangement might be worked out for sharing the costs of improvements.

Business support of an area occurs when the enterprise provides needed jobs, an enhanced tax base, and funding resources for improving roads, water systems, and so forth. The company passes these costs on to its clients and/or is able to secure some tax advantage from investing in these improvements.

RESOURCE STEWARDSHIP

Resource stewardship, so essential because of increased population, increased consumption, and a decreased pool of available resources, includes seven major components:

1. Efficient use of natural resources in products and in production processes
2. Elimination of negative environmental impacts on local natural resources, including air, land, and water
3. Creative use of "waste" resources and of industrial byproducts by entrepreneurs (e.g., plastics, rubber, paper, sulfur)
4. Recycling of used resources by consumers (e.g., oil, aluminum, glass, newspapers)
5. Search for new resources to meet present and future needs created by the depletion of current resources
6. Development of new, alternative uses for old abundant resources to meet human needs and relieve the strain of consumption on less abundant resources
7. Conservation of old and new resources, to keep the earth in balance and sustain earth's inhabitants into the future

GENERATIONAL IMPACTS

When a corporation locates its operations in whole or in part in a community, its impacts will endure whether its sojourn is brief or lengthy. These impacts include initial or subsequent plans for growth and expansion, and their implementation; the extent of local hiring for managerial, professional, and technical, as well as entry-level, positions; the positive and negative effects on local businesses: forcing closures or promoting growth; effects on local educational institutions that need to provide for employees' children; utilization of regional resources: natural, recyclable industrial, financial (which might preclude others' access), communication; waste generation and the effects of waste products on the environment and on the community's carrying capacity for their disposal; and health impacts, in the present (immediately apparent) and in the future (discernible only years or decades later).

Corporations and individuals have the responsibility to truthfully evaluate, where possible, such generational impacts. Some corporations fulfill this responsibility well, recognizing that they do not exist in isolation from the broader social community. Other corporations, unfortunately, neglect their social responsibility to the detriment of the local community and, ultimately, to their own detriment as well. Some individuals, too, avoid their responsibilities to the local and human communities of which they are a part: even when they are aware of harms caused by their activities, they do not act ethically because of their quest for financial gain or their concern about job security. Other people, by contrast, see the larger picture and either act through corporate channels or, if necessary, become whistle-blowers, agents for the public good who might suffer reprisals, such as job loss, demotion, or transfer, from industry or government. Such reprisals would cause financial hardship for such a person who places the public interest above personal benefit.

PUBLIC DOMAIN

Another way in which people work for the common good, even if it might result in personal financial disadvantage, is by putting their innovations into the public domain.

When inventions are placed by their creator into the public domain, no proprietary rights are assigned to them, so that anyone wishing to use them will have access to them free of charge. There are advantages and disadvantages to such practices.

The advantages of placing an invention in the public domain are that (a) everyone who might benefit from the invention is freely able to do so, needing to spend money only for materials (and possibly for labor, depending on their personal skills) in order to benefit from the invention; and (b) an invention that would be socially or individually beneficial is not kept from public awareness or use by a company that has obtained the rights to the invention to keep it from current production because it offers competition to a company product, or because its production will generate greater profits at a later date when a particular resource is depleted or when another product that fulfills the same function can no longer be produced. (The stunting of solar energy development in the United States is a case in point. Transnational corporations bought independent, innovative patent-holding small companies, thwarting solar competition for fossil fuel–based energy sources, and in speculation that the patents would be highly lucrative when world petroleum reserves were substantially diminished.)

The disadvantages of placing an invention in the public domain are that (a) people might never become aware of the invention and therefore never use it because of insufficient publicity about its existence; and (b) the invention might never be mass produced because the costs of its development and dissemination by a manufacturer would not be covered by profits that would be generated by exclusive rights to the innovation. In both cases, the social benefit intended by the inventor would not be achieved.

Commercialization, by contrast, although it runs the risk that a company might monopolize or allow only limited access to the invention, enables the invention to be used to benefit both the entrepreneur and society. Inventors wishing both to benefit society and to ensure utilization of their invention might choose a middle ground, granting entrepreneurs limited rights to their invention: limited use of aspects of the invention and a limited time frame in which to use them. In this way, a company could not "sit on" an invention solely to prevent its dissemination for public benefit.

COMMUNITY RELATIONS

If a company is to work well and avoid unnecessary intervention by government into its operations, good community relations are important. The purposes of formal community relations efforts should be:

- To discuss mutual understandings and expectations
- To promote benefits for both corporation and community

When the company and the community confer honestly about their respective requirements and their hopes for what might emerge from their relationship, and when they strive to find mutually satisfying means of realizing those requirements and hopes, good community relations result. For the company, this means that they will be a new "good neighbor" in town. For the town's citizens, this means they will contribute to and benefit from the satisfaction this neighbor derives from residence in this community, and from the economic, structural, and social enhancements the new neighbor brings to the community.

"Community relations," then, should not mean the type of "P.R.," or public relations, activities used by some companies that is intended not to inform but to deceive the local community about company operations, to divert community attention from pressing or potential problems, to stall for time in the presence of problems, to deny that problems exist, or to avoid taking responsibility for problems caused by the company. Actions of this sort would justify investigation and intervention by government (or by citizen groups when government is the culprit or is evading its responsibility for public health and safety) into company operations. Examples of needed intervention in the face of corporate or government neglect abound from the past twenty years—tobacco companies denying for decades that their products have been proved to cause major illnesses, despite studies affirming that fact by such organizations as the American Heart Association, the American Cancer Society, the American Lung Association, and independent laboratories, and then declaring that consumers know of possible risks and so are exercising their own "freedom" when choosing to smoke; the nuclear disasters at Three Mile Island on March 28, 1979, and at Chernobyl in April 1986, where governments at first denied and then minimized public health threats; the chemical disaster by Union Carbide in Bhopal, India, in December 1984; the chemical disaster at the Hoffmann-La Roche/Givaudan Icmesa plant in Seveso, Italy, on July 10, 1976; and the poisoning of air, land, and water by the U.S. government at the Hanford Nuclear Reservation in Washington State for decades, after the displacement of native peoples from their lands on the Columbia River, a radioactive pollution harmful to fish, native peoples, and windsurfers that was caused by weapons manufacture and denied by the government.

"Community relations" should not include such deception. Neither should it mean that a company maintains rapport with the community solely while the community concessions used to attract the company (for example, deferred taxation of the enterprise) are operative, and then moves production elsewhere to obtain higher profits. If a company seeks to move out when the incentives granted to it have been fully utilized, or in other cases where community well-being would be enhanced by public use of private property, the community might decide to exercise its legal right of eminent domain over private property.

EMINENT DOMAIN

The citizens as a whole, acting through their elected government officials, might decide that it is necessary to subordinate private rights to the common good through the exercise of eminent domain. This practice enables government, acting on behalf of its citizens, to take private property for needed public purposes, while appropriately compensating the owner suffering the loss of the property. The community must establish a compelling public need for taking private property. There must be a public health, safety, shelter, transportation, energy, or other economic or social benefit not otherwise reasonably attainable that outweighs the individual's personal or business needs and civil right to benefit from continued private use of the property.

PUBLIC–PRIVATE PARTNERSHIPS AND PROFITS

Private resources sometimes are insufficient to support the costs of research and development, while government agencies might lack the facilities and personnel to engage in developing and distributing socially useful products. In such instances, public–private partnerships can be established to facilitate research and development in such areas as medicine, agriculture, and computer and other scientific equipment. Often this research and development occurs in college or university settings, with research funds for salaries and equipment provided by government alone or by government and industry.

Government–industry partnerships can be very useful for developing technology which can be the basis for socially useful products and processes; unfortunately, the bulk of the research and development funding typically has gone for military purposes, with ultimate destructive rather than constructive intent.

Where socially beneficial technology results from these partnerships, the question of benefit beyond that to society emerges: Who should receive the financial gain from the resulting innovations? A

CONTENT IN CONTEXT 4.3

Eminent Domain

The Nifty Nail Company has decided to move its operations from Steele City to a foreign location. Although it has been in Steele for three decades and is making a profit each year, its owners have determined that they could make greater profits in Mexico because of substantially lower labor costs and minimal environmental safeguards. The Nifty employees are angry and fearful. They have worked hard for the company; their town is extremely dependent on Nifty for economic survival, and stores and schools would suffer. The city council decides to exercise its right of eminent domain over the Nifty plant, and run it as a city enterprise, arguing that the community good requires taking private property for public purposes.

The city of Flosse is concerned about the dental health of its children. After a presentation by a local dentists' association, the city council decides that it will increase taxes in order to add fluoride to the city water supply. The council declares eminent domain to remove the homes adjacent to the water plant so that it can be expanded to accommodate fluoridation equipment. Homeowners whose property would be seized question whether fluoride is safe for everyone to drink; assert that fluoride rinse for the few does not justify fluoride drinks for all; and object to losing their homes for the sake of others' teeth or profit.

Should either or both of the community needs cited above be considered sufficient grounds for city governments to declare eminent domain over private property? On what bases do you rest your judgment? What change in any of the circumstances described would alter your decision? What examples of eminent domain have you seen? Do you agree or disagree with them? Are there any examples of situations in which you think a government should declare eminent domain in order to meet a pressing public need with expropriated private property?

complaint of some analysts is that often government takes all or most of the risks, and industry takes all of the profits in some research and development (for example, new medical technologies). In other words, if nothing useful results from the investment, taxpayers absorb the costs; if something profitable emerges, the private company—or a university researcher suddenly establishing a private company— reaps the benefits of the jointly funded (or even solely government-funded) research, and resists paying back to government (the taxpaying public), from profits, the costs of research and development. Such a payback would, of course, in effect make the initial government funding a loan rather than a grant and would free up funds for additional government investments/loans in other partnerships in the future.

In order for public–private partnerships to continue to work well, to continue to be accepted by the taxpaying public, and to continue to provide useful products and social benefits, government must receive a reasonable, proportional return on its investment in them. Since government funds are much greater than private funds, and since government power and financial resources can counteract other nations' subsidies for private enterprises or use of state-owned enterprises for manufacturing and/or marketing goods (which gives public and private companies in those nations an unfair competitive advantage), public–private partnerships can prove to be very beneficial to the private sector and, ultimately, to their nation as well (in the form of jobs and financial well-being).

THE INNOVATION PROCESS

The innovation process begins with a creative reflection from within a particular historical context. The reflection is linked to prior discoveries, building upon or branching from them. The context is the locus of interaction between an individual, with her or his education and experience, abilities, and shortcomings; and the varied communities or social groups to which that individual is related, with all their complexities and contradictions, values and biases, insights and shortsightedness. The creative reflection might come from either of two orientations: community (people) as primary, or corporation (profits) as primary. These two primary development orientations for business are not necessarily exclusive: one might recall the corporate

CONTENT IN CONTEXT 4.4

Public–Private Partnerships

The Cosmic Rocket Company has used its advanced Data Control Corporation computer system to design a new type of engine that uses magnetic interactions—external to assist liftoff from earth and internal for travel in space—for interplanetary travel. The company applies to the federal government for development funds to concretize its theoretical concept. As CRC's president, Craye Cannon, explains in a meeting with government officials: "Our approach is innovative in the industry, technologically sound according to our own computer projections, will save enormous amounts of energy and money currently being expended on conventional rocket systems, and will make interplanetary travel quieter and safer without those big nuclear-waste-driven propulsion systems."

If the drive system becomes operational, Cosmic Rocket Company will hold the patent on a new product that will be safer, more fuel efficient, less expensive to manufacture, and faster than existing systems.

"Your idea sounds good," responds the Secretary of Energy, "but you're asking for a lot of public money for a private corporation's product and profit. What can the people expect in return? Granted, travelers will have a faster and better ride, but that's a small segment of the tax-paying public. What kind of interest on the loan, or profit sharing when the system is operational, do you propose to offer us?"

Should the government support Cosmic with this innovation? Why or why not? If so, to what extent? What revisions, restrictions, and reimbursements should the government require? What recompense should Cosmic request? What might be other possibilities for public–private partnerships? Do you believe such partnerships are in the public interest? Why or why not?

slogan of Control Data Corporation, "to address society's unmet needs as profitable business opportunities," or, expressed in a different but complementary way, "to address society's unmet needs while engaged in profitable business activities." Technological innovation might emerge through a company's consideration of its own commercial needs or its broader community's needs; meeting the former might result in meeting the latter, and vice versa. In either scenario, the corporation is able to make a reasonable profit while contributing to the overall well-being of the society in which it does business. In this way, both company and community benefit.

STEPS TO INNOVATION

Seven steps are part of the ideal technological innovation process:

1. Contextualization
2. Conceptualization
3. Communication
4. Cooperation
5. Concretization
6. Consociation
7. Commitment

These steps provide for an appropriate integration of corporation and community.

1. Contextualization

The corporation influences the community and the community influences the corporation. The interrelation and the interaction between the two can benefit both. The creative idea that comes to an innovator does not spring from a vacuum: it is based on the education and experiences not only of the innovator, but of the society from which she or he emerges and of other societies that have influenced the innovator's society. Each insight and each invention has a history prior to the moment it is expressed by the innovator. The innovator becomes the catalyst or the midwife, the agent through whom an insight or invention comes to light when the context and the moment are appropriate. The extent to which the innovation will be disseminated is dependent on the extent to which others—entrepreneurs, or government officials, or both—appreciate its significance and provide support for its development. Thus the community is both the foundation for innovation and the structure through which it develops.

The corporation does not develop and prosper in isolation. As it concretizes innovation, it provides, among other community benefits, employment; needed property taxes for community projects and programs; a product needed or wanted by society or by some of its members; and responsible working people, the foundation for a stable and prosperous society. The contextual relationship between corporation and community should be characterized by mutual respect and beneficial interaction:

$$\text{community} <> \text{corporation} = \text{contextualization}$$

2. Conceptualization

The innovative idea is conceived as a response to the context of its creator, and is given its first formulation. The innovator has analyzed a particular aspect of individual or social life and found a missing or better product to meet a need or potential want. The innovator formulates that product's components and development, and its relation to the situation in which he or she hopes it to be placed. Initial drawings and text are tentatively set down.

3. Communication

The innovative idea is explained by its creator to an entrepreneur who can obtain the capital and has the business drive to transform the idea into a concrete reality. (Sometimes the inventor and the entrepreneur are the same person.) The prospective business associates agree upon the potential uses and benefits (personal, social, and industrial, as appropriate) of the proposed product, and then begin to formulate their working relationship, including mutual responsibilities for product development. It is essential, at this stage, that the inventor and entrepreneur have an open and frank discussion about mutual responsibilities and expectations.

4. Cooperation

The inventor and entrepreneur work together to initiate the production process for realizing the idea. (Or, the inventor-entrepreneur organizes this process.) Mutually agreed-upon responsibilities (usually contractually expressed) are fulfilled. The necessary structures are built, equipment is purchased and put into place (including provision for maintaining a clean environment in the plant region), resources (natural, recyclable, and financial) are secured through contracts, employees are solicited, interviewed and hired.

5. Concretization

The innovation utilizes appropriate natural and, where feasible, recyclable resources, and conservation practices permeate production. These conservation practices include care of air, land, and water in the vicinity of the productive facilities; and utilization or recycling of byproducts of the production process. Plant managers carefully and continuously monitor emissions and effluents.

6. Consociation

Company and community work together to ensure company good and community good in the production and distribution of the new product. Company and community fulfill their respective responsibilities in their collaborative relationship: the company provides employment, pays its appropriate taxes, safeguards the health and safety of its employees and the local community, and conserves the environment in which it is operating; the community provides appropriate, agreed-upon services and infrastructural support such that the company might expect a good working environment for its operations. Mutual expectations are declared at this stage.

7. Commitment

Company and community pledge mutual support as successful marketing provides profits for the company and economic benefits to the community in the form of employment and of taxes supportive of community needs. When concessions have been made by the community to attract or retain the company, that company has at least a moral, and sometimes a legal, responsibility to maintain operations in the community as long as it is profitable to do so. "Profitability" is not considered solely in terms of "profit maximization": if that were the goal, a company making a reasonable profit in one community could pit that community against others regionally, nationally, or internationally in an unfair bidding war, with the casualties being wages, jobs, the environment, and local tax needs. If company owners and community leaders act responsibly for their respective goods, then a common good can emerge which enables sustainable economic benefits to both company and community, and catalyzes continuation of commitment by each party.

When these seven steps are followed, innovation provides a new product, company viability, job security, community vitality, and environmental sustainability. Real growth and progress take place in a local community and for the nation as a whole. The well-being of people and planet will be assured in proportion to the extent of regional, national, and global implementation of this process.

INNOVATION, CORPORATION, AND COMMUNITY

In the contemporary era, people suffer from oppression in the political and economic arenas and the earth suffers from exploitation in the environmental arena. The community (local, regional, and national) as a whole has a responsibility to promote justice for all of its individual members and survival of the commonweal. Individuals and communities are responsible for taking care of the land and resources entrusted to their care (in the form of private, communal, or public property). The corporation that is integrated into a community has a responsibility to the community, to its individual members, and for the land on which it is situated and the resources which it utilizes.

When individuals, communities, or corporations renege on their responsibilities all can suffer; when individuals, communities, or corporations become locked into only one approach to meeting individual, community, or corporate needs because "we've always done it that way," economic development will stagnate, employment will be lost, and the environment will suffer. By contrast, when individuals, communities, and corporations are open to an innovative process for meeting their respective needs, and work together to implement that process and integrate its components into the community setting, then technology, guided by communitarian and transformative ethical concerns, can offer both hope for beneficial and sustainable development and concrete practices to make that hope become a reality:

innovation + community + corporation
= social progress + environmental sustainability

In order for these efforts to be successful, people must be willing to experiment: to take reasonable risks when current ways of thinking and acting, and current social structures and business practices, are proving to be inadequate for promoting the good of individuals, communities, companies, and local ecologies. In this process toward true progress, old traditions need not be simply rejected and set aside: they might remain as the bases for careful consideration of proposed new forms of responsible human activity, and as correctives for overly idealistic projections of positive outcomes for new structures and practices.

ETHICAL PRINCIPLES

1. Companies and communities must work together to ensure both the company good and the common good when planning new ventures.

2. Entrepreneurs should fairly and honestly declare the expected social and environmental impacts of their proposed operations.

3. Plant sitings and operations must take into consideration local and regional social impacts.

4. Local and state governments must consider the long-term health and social impacts of industrial processes on their respective constituencies when attracting new business, allowing existing businesses to alter their productive processes, licensing new businesses, or considering imposing new constraints on potential or existing businesses.

5. Harms and benefits for future generations must be considered for resource extraction and consumption.

6. Social and environmental consequences of extraction, production, and consumption must be considered.

7. Plans for plant closings must consider, and propose realistic remedies for, negative impacts on the community.

8. Agents for public good should be rewarded, not penalized, for acting responsibly in situations where the public good is endangered or the public interest is threatened.

CHAPTER FIVE

Entrepreneurship, Employment, and Environment

Human communities and enterprises do not exist in isolation from other life communities and planetary forces, cycles, and rhythms. Technological developments utilize the earth's material resources and affect the earth's physical structure and ecologies.

People, no less than the earth's other inhabitants, have the right to use such of the earth's resources as are necessary to meet human needs. People also have the responsibility to acquire and use those resources in ways that do not threaten the long-term ability of the earth to sustain humanity and the earth's other inhabitants.

Sometimes consideration of those rights and responsibilities has led to debate over human needs versus human wants. At other times conflicts have arisen between entrepreneurs and environmentalists, with workers drawn in because of real or apparent (contrived) threats to employment and economic well-being. Responsible owners, workers, and environmentalists are learning that both employment and the environment can be protected through responsible and innovative appropriate economic development. In that process, human needs are met, jobs are more secure, business is profitable, and care for the earth is promoted.

Entrepreneurs will achieve their greatest economic and social success when the product they manufacture and market meets a human need or want, is produced under conditions that are safe and healthful for employees, uses well the resources it consumes for production, and has no harmful environmental impacts. Entrepreneurs in this circumstance will have their greatest economic success because consumers are interested in purchasing their product; employees achieve greater productivity when they are appropriately compensated and need fewer days off because of accident- or illness-induced leave; no cleanup of harmful effects of the production process is required; and no legal fees are expended on litigation against justified employee, citizen, or government complaints about plant operations. Entrepreneurs will have their greatest social success in this case because local citizens and their government representatives are appreciative of the economic benefits accruing to their community because of the enterprise, particularly when those economic benefits are not offset by harmful outputs from the production process or diminished by company attempts to renege on collaborative arrangements that gave it tax breaks or other incentives for locating in the community.

Employees will have their greatest job satisfaction when the conditions just cited are met, and also if they anticipate some long-term job security when working for a particular enterprise. Job security can be threatened by the loss of resources necessary for continued production, unanticipated harmful internal impacts from production, diminished demand for their product, or citizen complaints about the external effects of production on the community or region. In some of those cases job security and company profits can be restored through collaborative community–company efforts to remove factors that are impeding safe and viable production.

Environmentalists will be most content when a productive process, its product, and its byproducts are helpful to people and planet. Entrepreneurial activity will be accepted as appropriate if it will not unduly diminish nonrenewable resources, if it will carefully replenish renewable resources, and if it will cause no short-term or long-term harm to living organisms or to nonliving aspects of the environment.

When entrepreneurs, employees, and environmentalists work collaboratively, employment and environment become interactive and sustainable, and the earth as a whole benefits. To get this collaboration up and running, the three participating groups must cooperate in planning for the utilization of renewable and nonrenewable resources through reductive (extractive) and reproductive (regenerative) enterprises.

THE IMPACTS OF REDUCTIVE ENTERPRISES

Reductive or extractive enterprises are those ventures whose productive processes remove resources which can neither reproduce themselves nor be reproduced by human efforts. The diminution of nonrenewable resources by reductive enterprises irreversibly lessens the available pool of resources of particular types, causing temporary or permanent deprivation of particular products or benefits.

The impacts of reductive enterprises locally and beyond become an ethical concern when the earth or its biotic communities suffer or are threatened with suffering as a consequence of the enterprise's present or projected operations.

Air, Land, and Water

Communities located in the vicinity of reductive enterprises ordinarily profit from the presence of an industry that provides a local economic stimulus and population stability through the construction and use of taxable property and the provision of jobs for those who work on that property or from that property as their base of operations. But these benefits may be offset by harmful practices of industry. The harm could come through exploratory drilling into the soil, which might cause groundwater pollution; storage of the byproducts of the main productive effort, which might leach into the soil or be blown by the wind, affecting the quality of air, land, and water; smokestack emissions or sewer effluents; or severely altered landscapes.

Harmful impacts on air, land, and water could be diminished through appropriate industry monitoring of the productive process and, as necessary, taking steps to eliminate pollution. Sometimes the requisite steps might be costly. These costs can be recovered by passing them on to users, where feasible, or, as a last resort if it is in the general public interest, through government relief in the form of tax reductions or rebates.

Natural Resources

Reductive enterprises, since they decrease the available pool of natural resources, should be conservative in the use of those resources, efficient in developing them, and exploratory of possible alternatives to them.

CONTENT IN CONTEXT 5.1

Reductive Impacts

Monica Morton works for the giant Montana Mining and Munitions Corporation. In her work orders one Monday morning, she is directed to take her bulldozer to MMMC's newly opened Glittering Pyrite Mine to begin constructing a holding pond for cyanide from gold-mining operations. Monica arrives at the site with her dozer, and when planning her work realizes that the pond will be located above a pristine river that supplies drinking water to local communities and is a trout-spawning site. Monica has heard of local opposition to the new mining operation and is sure that the proposal has not yet been approved by government agencies. "The folks in town are sure going to be upset when they see this and realize what it means to their water supply. Some hoped for jobs and tax money for schools, but this could make their little community unlivable." Monica grew up in Montana and fondly remembered lazy summer afternoons when she and her dad went fishing in the Blackfoot River. "It sure seems a waste to hurt those trout. I wonder if we can move the pond site somewhere else near the mine?" Monica is looking around for an alternative site as her boss arrives.

What course of action should Monica take, and why? (1) Perform her assigned task; or (2) check with company officials to remind them of government requirements. If she opts for the first course of action, what responsibility will Monica have for environmental degradation and possible consequent community relocation? If she opts for the second course of action, what should Monica do if (a) her boss tells her that the cyanide won't hurt the fish or pollute the water supply for the town; (b) her boss tells her it is not her concern, and wonders if she has become an environmentalist; (c) her boss tells her not to worry and to begin work because the appropriate approval is forthcoming? What responsibility does Montana Mining and Munitions have to the local community and to the environment?

Life Communities, Regional and Global

The extraction and use of nonrenewable resources tends to be accompanied by types of pollution that extend well beyond the local community. Regional and even global effects might be felt as wind and water currents carry to distant places the residues of the production and utilization of those resources. In research and development, production, and utilization, industry should evaluate the potential and actual impacts of their enterprises on the biotic communities (living organisms in ecologically related clusters) likely to experience those effects and take appropriate measures to prevent or remedy harms to those communities. Local citizens who benefit from plant operations must have a sense of responsibility for the consequences of their operations, not solely consider local economic well-being. If every community thought only of the local advantages of their particular enterprises, harmful production impacts would increasingly be spread across the globe. Although a given community might benefit in the short term, eventually it would suffer from the harmful impacts of an operation set up to provide short-term benefits to another community. In contrast, if every community (and nation) were to eliminate harmful consequences of production through seeking and using alternative materials or processes, all would benefit.

THE IMPACTS OF REPRODUCTIVE ENTERPRISES

The impacts of reproductive enterprises impacts are not ordinarily as potentially harmful or exploitive of the environment as those of extractive industries. As their name implies, reproductive or regenerative enterprises are ventures whose operations include some provision for replacement or replenishment of the renewable resources they utilize. Ordinarily working in harmony with the earth's natural processes and resources, or using domesticated varieties of those resources, reproductive enterprises provide for replenishing what they have removed. They are not, however, always benign. The manufacture, use, and residues of agricultural chemicals used to enhance the growth of crops or prevent their decimation by pests, and of defoliant chemicals used in forestry to select for particular types of trees, for example, can harm people and planet.

Air, Land, and Water

Reproductive enterprises, when appropriate to the resource under production, can effectively restore air quality (reforestation of timber lands), replenish the land (organic agriculture), and renew the resource itself (tree planting for timber and habitat; seed planting for crops); and water can be conserved and, in some types of enterprises, recycled for additional use. Again, there is a qualifier to these ordinarily helpful effects of regenerative enterprises: their use of certain chemical inputs to assist production might endanger health in their region.

Natural Resources

Corporations working carefully with natural resources in reproductive enterprises can increase some resources and maintain others. The available pool of natural resources that thereby remains constant or is increased is able to continue to benefit (or, where new resources are discovered or new uses are discovered for already utilized resources, to begin to benefit) earth organisms, both human and nonhuman.

Life Communities, Regional and Global

The impact of reproductive enterprises on extended communities can be beneficial or harmful, depending on the practices used in production and distribution. In the timber industry, for example, animal habitat can be destroyed through timber harvesting but restored through new plantings. The potential for destruction is what causes conflicts between the timber industry and other users of the land such as hunters, hikers, outfitters, and campers. The fears of members of the latter groups are allayed when responsible forestry is promised, particularly if the company in question has a previous record of responsible use of lands under its control. The key to balance here is that there should exist alternative habitat for dislocated animals until reforestation occurs; otherwise there would be no animals to inhabit the renewed forest area. Industry biologists would have to understand habitat needs, ascertain that they are available in the region in the interim, and provide for their recovery in areas where trees are cut for timber, so that people and other life forms all could benefit from using present and future forests.

CONTENT IN CONTEXT 5.2

Reproductive Impacts

Paul Bunion is logging for Idaho Lumber Enterprises. As he plies his chainsaw, he rejoices that he is providing food for his family and work for local sawmills. Paul smiles when his dog Blue jumps from the tremor as the giant fir crashes to the forest floor. "Well, Blue, there's another thousand board feet for someone's home. That tree grew for a few hundred years, but we cut it down in a few minutes. That's progress!" Paul frowns momentarily as he recalls that his daughter, Virginia, questioned him the night before about a Dr. Seuss story, "The Lorax," in which the Onceler and his relatives had cut a virgin forest of unique Truffula trees until the trees were all gone. As a result, industrial and residential pollution had poisoned the air, land, and water, chasing away all the birds, animals, and fish. Paul resolves to speak to Virginia's teacher about confusing young minds with stories that undercut their parents' hard work. The story did remind Paul about some clear-cutting that he and other workers had done a month ago. "But our foreman, Slash Gordon, told us that ILE plants young trees in areas it clears of old-growth trees." Later, during his break, Paul hears Slash tell another logger that since this cut was well within the forest on private land, no one would notice if they did not replant; and that these trees were bound for Asia since ILE could make a higher profit on raw timber sold overseas than on timber sold to local mills. "Besides," Slash laughed, "we can blame the environmentalists for the loss of work at the sawmills."

What long-term possibilities will there be for work for Paul, the other loggers, and the sawmill operators and employees? What impacts does clear-cutting have? Should companies be required to replant private land? Should raw timber exports be restricted to protect local value-added jobs? Can loggers support their families if logging is restricted?

ORCHIDS, EAGLES, AND NATURAL RIGHTS

In a trek through a Canadian swamp in 1864, John Muir chanced upon a cluster of white orchids and was thrilled by this sudden encounter with rare flowers that were thriving unseen by human eyes. While reflecting on the incident, Muir concluded, contrary to the prevailing "wisdom" present in his day and ours, that nature does not exist solely for human benefit: earth's creatures have a right to exist for themselves and for their Creator. Millennia before Muir, the same concept was expressed by the author of the biblical book of Job, who has God saying to Job: "Is it at your command that the eagle mounts up and makes its nest on high? It lives on the rock and makes its home in the fastness of the rocky crag" (39:27–28). God speaks to Job, too, of mountain goats and hawks in the heights, and of fish and crocodiles in the depths, whose lives are unseen by people.

What might be drawn from the insights of Job and Muir is that all of earth's creatures—indeed, the earth itself—have certain natural rights, in the literal sense of the term, derived from either their status as creations of God (a religious perspective) or their status as equal coinhabitors of the earth (an ecological perspective).

In recent human history, the notion of "natural rights" usually has meant that individual citizens (or a particular group, usually a dominant or rising social class) have inherent rights derived not from human ordinances but from nature. These are "rights" whose existence is proposed as being recognized by reasonable people: life, liberty, and goods, for example, in the Western intellectual tradition. The older biblical understanding, and the insight presented by John Muir and others, however, provide a basis for examining the relationship of people to planet and of people to other residents of the planet.

It is obvious to the reflective person that people see very little of what transpires in the world about them, let alone what happens in the cosmos beyond their limited experience, electron microscopes and Hubble telescopes notwithstanding. From the micro to the macro, lives are lived and events occur that are not part of the human experience. Awareness of the immensity and intricacy of the cosmos might inspire humility and a certain sense of awe, calling into question the anthropocentrism so characteristic of some religious thought and some scientific theory and practice.

Traditionally, Western religious thought has had an anthropocentric tendency: humanity is at the pinnacle of creation, disobedient humanity has been redeemed by God becoming enfleshed in human form, people are managers or stewards of God's world. Similarly, recent scientific perspectives have that same propensity toward anthropocentrism: scientific research and development will lead to technological and medical advances that will enable humanity to "conquer" space, control the weather, alter rivers and forests to meet human needs and wants, and extend human life.

Alternatively, one might discern religiously that humanity is but a small part of creation; that God did not have to redeem the rest of creation; and that people's best relationship to God's world is not as managers but as respectful citizens seeking to live in harmony with others and with the natural (physical, biological, chemical) laws of the cosmos. And one might observe scientifically that space cannot be subjected to human control; that alterations to one aspect of the natural world can have harmful effects on another aspect; and that death is part of the process of being human—in other words, that *can* does not mean *ought* in the realm of scientific research and development, that the earth should not be a plaything for anthropocentric scientists any more than it should be a resource exploited by anthropocentric religionists.

As we look to the future prospects for the earth, in an era of excessive human greed and extensive human need, those who walk with the Spirit and those who work with science—and those at home in both contexts—should explore common ground for asserting and concretizing the natural rights of orchids and eagles, of other life forms, and of rivers, mountains, and winds as well. Whether these rights are seen as inherited from the Creator or as inherent in the cosmos, their recognition will lead to a renewal of the earth and to a restored interrelatedness of all who inhabit it.

The extension of natural rights to nonhuman entities will mean that a more careful consideration of their need to be sustained will have an effect on plans for economic development. Companies and communities will have to think about appropriate development strategies that are concerned about resource depletion in terms not only of the potential longevity of an enterprise but also of the probable survivability of a viable remnant of the entities impacted through resource utilization.

CONTENT IN CONTEXT 5.3

New Technologies and New Responsibilities

Dr. Mary Martini is doing research on a new electronic virus that would have the capability to embed itself into broadcast television programs to replace all nonsensical programming with a colorful kaleidoscope that would continue until an acceptable program came on. "People are tired of what passes for entertainment today," she reasons. "Television is destroying young minds and ruining our culture." Dr. Martini wants to experiment with her virus on her home television. She meets with Geralda Rimera, her local FCC official, and is told that government regulations require a stricter control environment. "After all," Rimera says, "you have no guarantee that the virus would limit itself to just your television. It might surf the air waves to other homes in the city and would have no natural enemies to contain it."

On her way home, Dr. Martini debates releasing "just one little virus" into her television. "What real harm could it do? Besides, if it were to escape, people would be grateful for a little artistic improvement brightening up their lives." She calls her next-door neighbor, Dr. Bowe Peepe, to seek her advice. Dr. Peepe sympathizes with Dr. Martini, since she has had similar problems with federal regulations. Dr. Peepe, who had been cloning sheep, wants to experiment with human cloning "to create a new world and benefit human society" but has been told that it is "out of the question."

What responsibilities do Dr. Martini and Dr. Peepe have to their neighbors and to their community? What government constraints should be imposed on private commercial research? Who should determine what risks to the broader community might be allowed for scientific advances that have the potential to benefit society: government or business? Who should define "sensible" programs or "beneficial" practices, and by what criteria?

NEW TECHNOLOGIES, NEW PRODUCTS, NEW RESPONSIBILITIES

In these last decades of the twentieth century, remarkable technologies have been imagined and constructed. The research and development process has led to new products that enhance human existence. There have been negative impacts as well, in terms of social displacement through corporate location and relocation; unemployment through replacement of labor-intensive by capital-intensive production; environmental degradation threatening life's quality and even life's existence.

Innovations, then, bring new responsibilities. Innovators and entrepreneurs, citizens and public officials must attempt to weigh the benefits and the harms of innovation. Technological progress must remain just that: progress. Imagination and creativity can meet societal needs and overcome societal harms, if corporation and community work together to make that happen. Today's byproducts, for example, could become tomorrow's resources through human ingenuity; and that same ingenuity can be exercised to find alternative resources and products, and appropriate technologies, to meet human and planetary needs now and in the future. The assumption of new responsibilities by companies and communities could negate the harms of the past, alleviate the problems of the present, and create a bright new future. Collaborative planning and action by companies and communities can foster that future.

INDIGENOUS PEOPLES — HUMAN RIGHTS IN HISTORICAL CONTEXTS

Indigenous peoples throughout the world have traditions, usually based on religious understandings, requiring respect for the earth and all its life forms. From the Gwich'in people living above the Arctic Circle to the Mapuche people living in Chile, and including all other native peoples north, south, east, and west in Europe, the Americas, Africa, Australia, and Asia, a reverence for the earth (often referred to as Mother Earth) and knowledge of how to live in relationship with the earth are inherent characteristics of the traditional life of the community.

As entrepreneurs enter into negotiations with indigenous peoples, they should be aware of and respect — and learn from — the environmental consciousness of representatives of those peoples who adhere to their traditions. (It needs to be noted that some westernized representatives of these peoples negate their heritage and are open to exploitation of native lands.) On this issue, entrepreneurs should not offer to poverty-stricken or invasion-threatened native peoples "economic development" in the form of manufacturing or energy plants that release harmful emissions and effluents; landfill sites that would absorb toxins from chemical, nuclear, or medical wastes; and gambling facilities that may provide much-needed income and employment to impoverished native individuals and nations but can also bring crime and the deterioration of traditional spiritual and moral values.

As indigenous peoples interact with the new technological, business-oriented, consumer culture, they should not have to choose between westernized survival and traditional values. In the past, their human rights have been violated in a variety of genocidal historical contexts; in the present, they should reacquire those rights for present and future generations, while teaching to other cultures the concept of respect for the environment that is part of their history. Entrepreneurship, employment, and environment can interact on indigenous lands among indigenous peoples, when respect for their rights and values is present. In that interaction, entrepreneurs might benefit not only by accepting ancient attitudes toward the earth, but also from appreciating ancient appropriate technologies that work with the earth.

SUSTAINABILITY

In the area of economic development, sustainability has different meanings.

For an entrepreneur, sustainability might mean the potential for a company's long-term viability in terms of its access to financial, natural, and human resources, and its ability to integrate these resources in such a way as to ensure ongoing profitability.

For a community, sustainability has a broader meaning. It includes the viability not only of a particular company but of all enterprises within its jurisdiction or area, and the potential for long-term benefits not only from company operations but also from available area resources and the metabolic processes of separation and synthesis inherent in any dynamic setting. Because a community has to be concerned about the well-being of its citizens as resi-

CONTENT IN CONTEXT 5.4

A Sustainable Enterprise?

The Filly Firefight Company of Philadelphia has developed the Filly Forty-four, a handgun for women. The Filly pistol is lightweight plastic, fires plastic bullets, and is a new compact design that fits handily in purse or pocket. Excited about the new product, Filly CEO Bob Bullet travels south to find an appropriate community for a new Filly munitions plant. He speaks to the city council of Friendly Fire, Florida. "We'll provide needed jobs in Friendly Fire, and this is a real growth industry. More and more people are buying guns to protect themselves, and with crime and the population both increasing, we'll be able to work together for generations." Mayor Pharr Wright agrees: "We need jobs. Also, we've had plans for a new industrial park and this plant could be its anchor." Laura Leffte on the city council disagrees: "Let's hope that Congress will pass its proposed gun control legislation and slow down crime. It includes a ban on plastic guns because they can't be seen by airport metal detectors and are a preferred weapon for terrorists for skyjacking. For this industry to be sustainable, we have to hope that the crime rate stays the same or increases, something our constituents certainly don't want." Mayor Wright counters: "The Second Amendment guarantees to citizens the right to bear arms, so even if crime decreases, guns will still be sought by people. Besides, if we ban guns, only criminals will have guns. And anyway, guns don't kill people, people kill people." Leffte replies: "Actually, bullets kill people, so maybe we should ban bullets—no offense, Bob—instead of banning guns or people. And the Second Amendment links the right to arms to service in a state militia. It's not an unqualified individual right. It's linked to community well-being."

As the third member of the city council, how will you cast the deciding vote? What will be you reasoning, since you want jobs but also want sustainable economic growth?

dents of and as workers in a local or regional area, community representatives must analyze what is brought into the community as well as what is exported from it: for example, they must evaluate exchanges such as the introduction of hazardous substances to support a manufacturing process versus the jobs and tax revenues that the process provides; the pollution potential of emissions and effluents resulting from that manufacturing process; and consumer products brought in (such as lawn chemicals), and storm drainage flows sent out, by area residents.

For a scientist seeking objectively to study relationships within a region, sustainability might mean the capacity of an ecologically integrated area to maintain a balance among the various life forms and dynamic processes inherent to the region, including the ability of components of the region to adapt to each other and to adapt to or absorb extraneous impacts upon the area (for example, earthquakes or uncharacteristic violent storms); intrusions into the area (for example, migrating life forms or regional human industrial operations); or inputs in the area (for example, agricultural practices such as the installation and use by local farmers of new irrigation systems, or the arrival of acid rain or of pollutants from other areas). Here, too, there is a concern for metabolic processes as impacts, intrusions, and inputs occur: can the region accept or must it reject changes, and if it must reject them in its own interest, does it have the authority to do so?

For a theistic citizen from the perspective of religious faith, sustainability might mean the degree to which the earth is cared for consonant with its ability to exist as part of God's creation; its capability to maintain existential harmony among the natural laws of biology, chemistry, and physics to ensure its survivability and the survival of its life forms, its rhythms, its inherent essence, and its interactive components; and its capacity to provide for the competing or complementary needs of its inhabitants.

All of these perspectives on sustainability have in common a sense of the long-term relationships inherent in any context, relationships among different types of organisms and between those organisms and the soil, climate, and hydrology of their specific regional context.

In its most comprehensive sense, then, *sustainability* might be defined as the capability of a defined geographic area to maintain the conditions and resources necessary for the existential requirements of its various and diverse members (life forms) and components (such as minerals, climate, soil type, hydrology, magnetic fields, molecular combinations) in their particular existence-maintaining or existence-enhancing activities, and to integrate those particular activities in a balance with the existence-maintaining or existence-enhancing activities of other components or members indigenous to the area or incoming to the area.

Obviously, the different members of an area might have distinct and sometimes conflicting ideas or instincts about their particular existential requirements. Conflicts might be resolved by the domination of one member over another, by the mediation of an objective outside observer, or by the biological, chemical, or physical processes inherent in the laws governing the elements and compounds of the area as activities carried on within its boundaries are coherent with or contradictory to those laws.

THE PROCESS FOR RESPONSIBLE TECHNOLOGICAL DEVELOPMENT

Technological development is appropriate for a corporation and a community when it uses regional resources more than imported resources to the greatest possible extent and in the best possible way, when it has long-term viability, when it relates corporation and community well, and when it does all of these with overall positive environmental impacts. Appropriate technological development takes place when the community context of development—in terms of both people and the environment—is analyzed by local citizens who want to take responsibility for their own economic and environmental future, and who integrate regional resources, technological innovation, and environmental concern.

STEPS FOR APPROPRIATE TECHNOLOGICAL DEVELOPMENT

There are seven steps that can lead to appropriate technological development for a community, a region, or even a nation:

1. Values analysis
2. Natural resources analysis

3. Recyclable resources analysis
4. Human resources analysis
5. Outreach resources analysis
6. Financial resources analysis
7. Market analysis

If a community participates in efforts to take these steps, and if a group agrees upon the prioritization of community needs and community projects, the development that ensues will truly be "progress" for that community. Sustainable enterprises will be developed that integrate well the concerns of ethics, the reality of economics, and the long-term needs of the local environment.

1. Values Analysis

Prior to or concomitant with community development planning, citizens should decide together on the values they consider the most important to their community. There will be some values that differ, but there will also be shared values that can help to unite the community in the development process. These values might surface and be agreed upon by consensus when community discussion focuses on these questions:

- What are the operative values of individuals and organized groups in the community?
- Are some values in conflict?
- Could conflicts be resolved through reasonable prioritization of values?
- When prioritization is complete, which individual and group values overlap?
- How might these values guide this community's development?

2. Natural Resources Analysis

As the community seeks to incorporate its values in development efforts, it begins to analyze the resources available to it for development. First, local and regional natural resources are studied. These resources might be known because of past economic activity; others might be discovered through careful analysis of the elements, terrain, and subsurface materials of the area in which the community is located. Resources are listed and, optimally, mapped on a computer program that can be updated as new resources are discovered or old resources are depleted.

In this stage of the process, citizens would ask:

- What are the locally and regionally available natural resources: plant, soil, water, wind, solar, mineral?
- Will they be utilized through reductive enterprises, reproductive enterprises, or both types of enterprises?

3. Recyclable Resources Analysis

In Western industrialized societies, individuals and businesses have become accustomed to using a product and discarding the materials which brought that product to the moment of use. Consumers, for example, use prepackaged foods and throw away the material in which the food is encased: cans, bottles, boxes, and so forth. Companies manufacture wood products and discard the sawdust and wood chips. Publishers produce a daily newspaper and throw away the printing plates. Chemical plants produce a household cleanser but reject as waste other chemicals not incorporated into the cleanser. As landfills become saturated and streams become polluted, manufacturers, governments, and private citizens are seeking relief from the mountains and rivers of waste produced by affluent societies consuming their resources as if these were unlimited. Innovators are exploring ways of recycling some materials previously regarded as "waste" or "byproducts," transforming them into marketable products. For example, extremely durable window and door frames are constructed from wood chips combined with plastic bottles, wood pellets for home heating from compressed sawdust, and solar collectors from printing plates and burned-out fluorescent light bulbs.

Members of the community should explore these questions as they look around for recyclable resources:

- What materials are being discarded locally and regionally from current production?
- How might byproducts be used instead of being wasted?
- What byproducts might be utilized from projected production processes because uses for them have been developed independently elsewhere or complementarily during production planning?

4. Human Resources Analysis

An important resource sometimes overlooked in plans for sustainable development is the diversity of work skills available to a local community. In this aspect of local community analysis, citizens list the competencies present in the community in order to have an employee basis, when the other resource analyses are done, for expanding existing businesses or inviting new businesses into the area.

The following questions will help establish the local employability factor:

- What skills are locally and regionally available: basic skills, artisan skills, communication skills, managerial skills?
- How do these skills correspond to other available resources?

5. Outreach Resources Analysis: Communications and Transportation

In order for a product to be marketed successfully, communications capabilities and transportation facilities must be appropriate to that product. In today's era of enhanced telecommunications, computer modems, and faxes, the product might be ideas or services as well as a particular hard good. Consequently, communities must evaluate their ability to deliver a product—goods or services or both—to interested consumers or companies.

The determination of available outreach resources will emerge through these questions:

- What communications media are in place or might be developed?
- What modes of transportation are available, underutilized, and capable of enhancement?
- What modes of transportation might be developed, considering realistically the community's location, terrain, and sustainable environment requirements?

6. Financial Resources Analysis

Communities might dream of a variety of sustainable development activities, given the availability of the types of resources already mentioned, but these dreams will not come true if there are no financial resources forthcoming for development. Therefore, citizens must explore realistically the potential of various types of enterprises to attract capital. If the community has carefully mapped its other resources and realistically projected a means to develop them, the likelihood of attracting capital is enhanced. Communities should not limit themselves to seeking capital from one source or one type of source: there might be available, depending on the type of development proposed, a combination of public- and private-sector funds for the development of small businesses or major manufacturing facilities.

Community members familiar with financial matters should explore answers to the questions:

- What types of financial support are or might be available from private entrepreneurs, investors, nonprofit organizations, foundations, local, state or federal government, or international agencies?
- What are the conditions required by these sources for investment, for repayment as appropriate, and for continued support?
- How would the community and individual companies meet those conditions in both the short and the long term?
- What individual and community commitments can be made to secure necessary financing for new or expanded enterprises?

7. Market Analysis

While the community is drawing up its plans for sustainable development, it must ask, as various possibilities for goods and services emerge, whether there is or will be an interest in the various products possible of production in its locale. The market analysis should be realistic so that people's expectations will not be raised unduly and so that a long-range viability might be determined for specific goods and services capable of production. This market analysis should also seek to ascertain the extent to which there will be a demand for this product in the present and in the future, so as to determine scales of production. The community should remember that even the best of ideas might not be marketable if there is insufficient interest in the broader society for the innovation developed by the community as a new product, or for additional quantities of existing products.

Key questions in this type of analysis would include:
- Is there a need (secondarily: a want) that products capable of development might meet?
- Is that need able to be met consistently over the long haul, given other community resources?
- Is that need on a scale sufficient to sustain the projected enterprise(s)?

When these seven steps to sustainable development are followed, corporations and communities can devise collaborative efforts that facilitate:
- economic viability for the company
- economic stability for the community
- employment opportunities for members of the community
- environmental sustainability for company and community

ETHICAL PRINCIPLES

1. Individuals, businesses, and nations must respect the basic needs and inherent rights of biotic communities.

2. Production processes must take into consideration health impacts on local, regional, and global biotic communities.

3. Industry must design, construct, and operate safe and healthy production facilities.

4. Industry must immediately warn potentially or actually impacted communities of hazardous accidents.

5. Industry must take responsibility for hazardous accidents.

6. Industry must justly compensate victims of hazardous accidents for life, health, and employment impacts.

7. Industry must restore environmental quality after extraction of resources, after industrial accidents, and after intentional pollution.

CHAPTER SIX

Spirit, Science, and Society

The human spirit seeks more than merely material satisfaction in the pursuit of technological progress. While technological advances provide positive physical and social benefits for individuals and communities, including the individuals and companies that have developed them, there are also, potentially, spiritual benefits that provide a holistic sense of well-being for their developers and users. Such benefits include a sense of having utilized well the resources at one's command, having made a positive contribution to the human enterprise, and having conserved the well-being of the planet. Such values are seen as beneficial in themselves (the secular tradition) or in relation to the Creator (the sacred tradition).

SOCIETY'S ROLE IN PROMOTING SOCIAL VALUES

The organizational or institutional structures of societies provide ideas or policies useful for promoting social values. These institutions might be educational, governmental, political, financial, or religious. They might be pursuing their own interests and include the social good secondarily or unconsciously, or they might be seeking the social good primarily and consciously, or integrally and consciously, as part of their philosophy and practices. Similarly, individual citizens might perceive personal benefit from meeting societal requirements, or they might make personal commitments to the general good. In both cases the social good is more likely to be accepted and achieved.

Self-Interest

Self-interest can serve to promote social values when:

- individual values coincide with social values
- individual needs coincide with social needs
- individual actions are informed by a certain sense of reciprocity

The first two conditions for actuation of social values are difficult to project because of some people's tendency to seek first what most serves their own needs and wants. The third probably occurs more frequently, and is sometimes expressed as the Golden Rule: "Do unto others as you would have others do unto you." The individual acts in a certain way in the expectation of being treated similarly by others in the future.

Those who expect appeals to self-interest to remedy social and environmental ills usually believe that the person would be well enough informed to know what is in their own ultimate self-interest. But oftentimes people do not reflect carefully enough and, even when careful reflection occurs, still do not act as they believe they ought: an immediate, almost unreflective, act in pursuit of perceived self-interest, often based on a quest for some short-term gratification, overrides any sense of long-term responsibility for self or community. In other cases individuals might thoughtfully and deliberately choose a course of action harmful to the social welfare because their own perceived self-interest comes into conflict with the needs of others or with their responsibilities toward others and to the world as a whole. They act selfishly from deliberate calculation.

Enlightened Self-Interest

People sometimes are willing to forgo short-term personal benefits in order to provide for or promote the long-term interests of the community as a whole. They might act in this way out of a willingness to place others' needs above their own wants or even needs; or they might be acting from enlightened self-interest, identifying their well-being with community well-being. Individuals are acting from enlightened self-interest when:

- they believe that they will benefit as a member of the community, experiencing the effects of their choice in an indirect way; or
- they believe that their own long-term interest is entwined with that of the community; or

- they believe that they, too, might in some way directly benefit in the long term from a choice made in the short term in favor of the community.

Enlightened self-interest can promote responsible technological development when people are perceptive enough to see long-term benefits for themselves and for the community as having greater value than short-term personal gratification. When people act according to these considerations, benefits will accrue to the community and to the earth not only in the present, but also over generations.

Self-Motivation

An ideal way to foster the social good is through self-motivation. It is ideal because individuals with this perspective recognize their responsibility for the commonweal and act to fulfill that responsibility. In fact, some individuals might go beyond responsibility to self-sacrifice: they are willing to do more than might be required or expected of them in order to assist a particular person or persons, or to aid the community as a whole.

Factors inhibiting substantial and sustainable achievement of the social good through self-motivation include the following:

- the limited number of such individuals, and the possibility of their forceful elimination from the community
- loss of energy, interest, or commitment on the part of the individual (sometimes called "burnout")
- the extent of resistance that self-motivated persons encounter in the broader social arena to their and others' efforts at social reform or transformation

When these inhibiting factors are pervasive in a particular community or culture, it is difficult for most people to be self-motivated to act on behalf of their community.

Community Motivation

Another ideal way to foster the social good is through the commitment of the community as a whole (city, state, or nation) to seek, strive for, and satisfy the needs of all its members. When this community motivation is present, impediments to community betterment can be overcome.

Impediments to substantial and sustainable community orientation toward the social good include:

- changes in community membership
- loss of inspired and compassionate community leadership
- alteration of the community perspective from social concern to individualistic interests

These impediments to orientation toward the social good might be offset by other operative community factors that promote community endeavors to meet the needs of all community members. These factors are:

- community stability, which might be provided by community efforts to promote sustainable economic development
- dynamic, committed, and concerned leadership in the community, not only in the work of community-oriented political officials but in staff and citizen work to promote a viable community
- mutual concern within the community for all its members, regardless of their economic class, ethnic group, and so on, such that compassion for those in need—a broad social concern—takes precedence over individual selfishness

Community motivation can help to eliminate divisiveness and bring about relatedness, and can promote community sharing, community prosperity and community viability, and community pride and community happiness.

Corporate Policy

Corporate policy promotes social values when the company has a good working relationship with the community (or communities) in which it is located, when it actively pursues the good of that area, or when it seeks to meet the area's pressing needs by providing new products, services, employment, clinics, housing, and so forth, or by encouraging voluntary community involvement by its employees.

Corporate policies to promote the well-being of the community in which the corporation is situated include provision of released time

CONTENT IN CONTEXT 6.1

Government Coercion

Donald Shays is a farmer whose family has worked the same land for three generations. He has a diversified operation: dairy cows and grain. One Monday, when he walks with his dog Rebel down the lane to the highway to check his mailbox, he finds a letter from the county government planner stating that the city of Eden Vale needs to take his land for a sanitary landfill, but will pay him a "fair price" for it. The city dump is nearing capacity, and his farm is within reasonable driving distance of the city limits but far enough away from the city that residents would not be offended by the odor from the new dump site. Donald is upset about the county government's proposal, and tells Rebel: "This is good, productive farmland, and a good home for Danielle and our children. Why do they want to take it out of production and turn it into a garbage dump?" When he returns to his house he wonders to his wife, "Why do city folk generate so much garbage? It's their problem, not ours. Why don't they just bury it in their own backyards? It's their garbage and their problem." Danielle observes that she saw a newsmagazine article discussing "sacrifice areas" selected by the federal government for toxic waste dumping, and wonders: "Can the county designate 'sacrifice areas' for garbage?" Donald and Danielle Shays decide to write a protest letter, get a lawyer if necessary, and propose to city officials that (*a*) they use less productive land for a dump; and (*b*) they begin a recycling program so that future landfills will not fill as quickly.

What resolution might be made in this conflict between the social good (sanitary disposal of garbage) and the individual good (retention of productive private property, home, and employment)? Are there some circumstances in which you would agree with the city government, and others in which you would agree with the farmer?

for employees to work as community development advocates while salaried by the corporation; encouragement or even expectation of community involvement by employees; some form of corporate recognition for services contributed by employees to the community; and corporate matching funds for community contributions made by employees.

GOVERNMENTAL AUTHORITY

A major function of government is to promote the common good. This role implies that government will restrict efforts by individuals or groups within its jurisdiction to oppress other individuals or groups, particularly where people's fundamental needs are not being met, or where fundamental rights are being denied. When government fulfills this responsibility, it promotes community stability and individual well-being and fosters the economic and political viability of society.

The social good should be pursued by government coercion—through civil laws, police activity, and the courts—as a last resort in contexts where self-motivation and community motivation are lacking, and efforts at moral suasion fail. It would indeed be wonderful if nations could have less government, as envisioned by Thomas Jefferson, Henry David Thoreau, and Karl Marx. Unfortunately, social constraints are necessary to protect society from the actions of people motivated by greed for money or power. These constraints are expressed through laws passed by legislative bodies and enforced by the police and the judicial system. Although most people resent and sometimes resist government's use of its coercive power, in principle they accept it when the system acts to arrest and imprison dangerous criminals, or uses collective tax monies to pave streets, build schools, and care for the elderly. Government authority, again, should be used in a coercive way only in the interest of promoting a true community good, and always taking into account the needs and rights of those who suffer some loss due to government's exercise of this extraordinary power.

DIVINE SANCTION

Some citizens believe in divine interest in social good, and engage in efforts to promote the social welfare not only through sharing their religious teachings and offering their prayers, but also through active community work on behalf of society's downtrodden. This work might take the form of acts of charity, community organization, or even efforts at profound community transformation. Religious institutions can have a powerful impact on efforts to promote the common good. Religious values permeate the laws of many nations and usually have become so embedded that their influence is not recognized (except in nations having an official religion). Religious institutions are also, even when not attempting to be so, political institutions: their advocacy of human rights or workers' rights, for example, even when not done through formal lobbying efforts, can have far-reaching political impacts in nations and in the international community.

In some circumstances, religious institutions might be detrimental to social progress or to harmonious human interaction. They might reject any appreciation of the divine other than their own. They might absolutize one particular interpretation of even their own religion. They might produce a leader who is a demagogue. A particular historical era might be absolutized, and so science and technology might be rejected. Followers of a religion can become fanatical about their belief system: there can be serious consequences in a regional, national, or international context when devotees believe that God inspires and approves the actions only of their religion's faithful or of the most fervent of these faithful.

For better or for worse, then, divine sanction can have an important influence upon the development of business and technology, and upon the ethics that guide that development. The influence and the ethics will be positive to the extent that divine sanction is utilized to promote respect and compassion among peoples, and human responsibility for society and sustainability.

SPIRITUAL INSIGHTS FOR AFFIRMING TECHNOLOGY AND NATURE

Insights for fostering social values come from both intellectual and intuitive or inspirational reflections of people evaluating community needs, or pondering human responsibility to a greater Spirit. These insights can come from humanist and sacred traditions, and from secular or spiritual sources and thinkers.

CONTENT IN CONTEXT 6.2

Church and State

Reverend Daniel Korahs in his fiery Sunday sermons taught his congregation in Paradise Grove that only Christians worship the true God, and that God favors "the good old U.S. of A." over all other countries. Appalled by some of the new textbooks used in his daughter's public high school, he convinces two members of his congregation to run with him for election to the school board, and the three are successful. At the first meeting Daniel and his colleagues, a new majority on the board, move and pass several policy changes: each class should begin with the Lord's Prayer; only the "classics" of Western culture should be read, not the newer multicultural anthologies of literature; creationism should replace evolutionism in science classes; and individualism and the economic system should be praised for having created a wealthy nation, and not questioned in any way. "After all," Korahs declares from his position as the new chairman of the board, "our American culture is the best in the world, and our Christian heritage and economic system made it that way!" One of the mothers at the meeting respectfully notes that her family is Muslim and that while she respects Christian beliefs—"Jesus is mentioned in our Qur'an as a great teacher too"—she doesn't feel it's right to require all children to pray to God only in the way taught by one religion.

What might be the impacts of Reverend Korahs's actions? Since in his country "the majority rules," must the other citizens follow his rules until the next election? How might children and parents of other religious faiths view and respond to the new policies? Does religion have any place at all in a public school system? What values do children learn or not learn when religious ideals are absent from the body of culture transmitted in the schools? What might be the positive and negative societal impact over generations if education is declared to be "value free"?

Humanist Traditions

Over the past several centuries in the Western world, traditions have developed whose focus is less on God and whose locus is more outside religion. These humanist traditions have a greater tendency toward anthropocentrism (human-centeredness; human preeminence; human primacy) than religious traditions have. Whereas religious traditions have had a secondary anthropocentric emphasis (a view of God or gods as especially favorable toward human needs), the newer traditions have a primary anthropocentric emphasis: divinity is no longer necessary, humanity will strike out on its own and conquer new worlds of physical and social science, and humanity will bring freer inquiry to bear on a material basis for human happiness and well-being.

The humanist focus catalyzed numerous technical and social advances for humanity, but, by removing God from the picture, gave humanity greater liberty to exploit the earth without a sense of responsibility to a higher being and without fear of retribution for violating stewardship requirements demanded by that being. Human "progress," anthropocentrically defined, supplanted the perceived religious justification for "subduing the earth," with consequences no less devastating than those resulting from supposed divine approbation for planetary despoliation.

In the twentieth century, however, humanist thought became, in some instances, preoccupied with the earth-destructive consequences of scientism linked to individualism in capitalism. Humanist thinkers and actors began to strive to restrain human technologies and human activities which were environmentally threatening. In philosophy, literature, the social sciences, and the physical sciences people began to suggest alternative modes of human interaction with the earth.

The result, in recent decades, has been humanist calls for conservation, based on a scientific and philosophical appreciation of earth's fragility and fecundity, akin to religious thought through the ages that called for stewardship of God's creation. These calls have sometimes appealed to human self-interest (care for the earth or we will perish, or we will not find cures for illnesses, or we will have no place to go to appreciate natural aesthetics, and so forth), but at other times have focused on the inherent rights of other creatures and even of the earth to such an extent, for some, that other life forms are viewed as equal to humans, and that the survival of the earth is given higher priority than human survival.

Humanist thought, then, has become less myopic and optimistic, and more perceptive and reflective, about the consequences of unrestrained science and technology. Humanists' preoccupation with human injustice and environmental degradation stimulated a more critical humanist evaluation of technological development. Humanists began more thoroughly to evaluate innovations not only in terms of their potential profits for entrepreneurs and their perceived benefits to humanity, but also in terms of the totality of their consequences—both helpful and harmful—for the human family in particular and for the earth's biotic community as a whole.

The anthropocentric tendency of humanism became tempered, therefore, with an appreciation not only of human ingenuity but also of human dignity and of human interrelatedness with the environment. In other words, even while sometimes rejecting religion, humanism still came to appreciate values present in traditional religious thought: compassion toward the neighbor and concern for creation. Humanists came to evaluate technology as the concretization of human inventiveness within a context of concern for people and for the planet.

The humanist tradition, therefore, now offers—to the atheist or agnostic in particular, but also to the believer independent of institutional religion—perspectives on responsibility to the earth that parallel the best of religious traditions' admonitions to people to care for creation on behalf of the Creator. In this regard, fundamentalist assertions to the contrary notwithstanding, the humanist tradition approaches sacred traditions in environmental perspectives, and complements sacred traditions in environmental practices.

Sacred Traditions

In sacred traditions, God or the Spirit (addressed by many different names) is the One who envisions universes and their relations, formulates the laws that both bond them and catalyze their evolution at appropriate moments, gives birth to them, and continues as Presence to give them life and guidance. Spirit is both transcendent and immanent: One who is and One who becomes, the Thinker and the unfolding Thought, a being independent of creation and a being permeating creation.

The Spirit by definition cannot be quantified, identified, or analyzed: Spirit is of a different order of being and of reality than matter and energy as we understand them. This reality frustrates those scientists who want to measure, and categorize, and explain all that is; since some believe that all being can be so reduced, they deny any independent reality to Spirit. Those who have encountered Spirit are not bothered by those scientific attempts except as they harm others and harm the earth; those who believe in Spirit on the basis of faith in others' testimony sometimes are troubled by such scientific endeavors, and might even react violently to them.

Spirituality is the experience of Spirit by individuals and communities, an experience that comes from the self-revelation of Spirit both to those seeking to understand and approach Spirit and to those who have an openness to Spirit, whether expressed or unexpressed. Spirituality is both the most personal and the most universal experience of Spirit: it is an individual openness and response. Unlike religion, which emerges from and is institutionalized in a particular culture, spirituality transcends efforts at institutionalization and cultural restriction, and links all people and peoples.

Religion is the formalization of spiritual experience. The oral or written religious tradition of a particular people is their effort to describe, through the limited medium of language, the spiritual experiences of that people as a whole or of an individual or individuals within that people. The religious structure of a people is an effort to institutionalize and reenact those experiences to benefit or control the people who have not otherwise participated in them, and to guide or control the theory and practice of commemorations of, or reflections on, those experiences. Religious ritual is the ceremonial formalization of spiritual experience. It serves to commemorate individual or communal spiritual moments; to enable those who lack spiritual experience or who lack openness to their own spiritual possibilities to participate at least vicariously in others' experiences; to catalyze spirituality; to strengthen spirituality; to inspire people to join, to remain faithful to, or to fulfill their responsibilities in a particular faith; and to foster community among spiritual individuals.

Religions originate in particular historical contexts and organize particular historical constructs. They emerge from particular cultures at particular moments as those cultures' formalizations of spiritual experience, founded initially on people's time-based perceptions of the divine and expressed in terms appropriate to their culture. They may or may not have a sense of process and an openness to new experiences and new understandings. Without that process and openness a religion can become anachronistic, fanatical, and imperial, negating the spirituality and Spirit Presence in other contexts and cultures, the ongoing guidance of Spirit in its own culture, and the creative, imaginative, and envisioning use of human reason when people image God.

In their approaches to the world of nature, religions might teach:

- cooperation with the earth
- conflict with the earth
- a combination of both relationships

The primary focus for a religion will depend on its understandings of the Creator, of creation, and of human relationships with and responsibilities toward both Creator and creation.

In their approaches to technological innovation, religions might:

- reject technological advances outright
- embrace technological advances unquestioningly
- carefully weigh the projected social and environmental impacts of proposed innovations

Sometimes, religious insights brought to bear on proposed (or actual) technology can bring a new perspective that enables the scientific community to evaluate more realistically and holistically the projected effects of particular technological proposals or projects.

Biblical Traditions

In biblical traditions, the cosmos is seen as the handiwork of God. What God has created is good. People as images of God are to work creatively with the earth to meet human needs. The earth is entrusted to their care that they might do so, but the earth is not created solely for that purpose, nor are the other creatures inhabiting the earth, nor are the other goods provided by the earth. While people alone are God's images, other life forms have an inherent dignity which is to be respected. The Earth, too, is not to be carelessly exploited but carefully cultivated.

The biblical perspective of care for creation is present in both the Hebrew Scriptures (Old Testament) and the Christian Scriptures (New Testament). In the Old Testament or Hebrew Scriptures we read in Genesis (2:15) that "God took the man [symbolizing all people] and put him in the garden of Eden [symbolizing the earth] to till it and keep it." In Leviticus (25:1–7), the sabbath year is proclaimed, which requires a year of rest for the land every seven years in order that the earth's soil might be rejuvenated; and in (25:8ff) the jubilee year is ordained, which requires land reform and redistribution every fifty years, so that the earth's landless people might renew their relationship with the soil. In Job (38–41) the beauty and inherent dignity of all of creation is affirmed, from burning stars through inanimate stone and to animals dwelling on mountaintops and ocean depths: the natural world is blessed by and blesses God in images such as these poetic descriptions of mountain birds (39:26–29): "Is it by your wisdom that the hawk soars, and spreads its wings toward the south? Is it at your command that the eagle mounts up and makes its nest on high? It lives on the rock and makes its home in the fastness of the rocky crag. From there it spies the prey; its eyes see it from far away." (Note that these birds are not on earth to benefit humanity: they have an inherent dignity of their own.) In Psalm 148 this thought is continued, where all creation is called upon to praise God. In the New Testament or Christian Scriptures we learn in the Gospel of John that "the Word became flesh and lived among us" (1:14), an incarnation that affirms the worth of the material world because God embraces it in a new way. In all the Gospels the Christian experiences of the risen Christ are expressed, affirming again the material aspect of people: Jesus is not just immortal (which would emphasize the primacy of eternal spirit over mere matter), he is resurrected (which affirms, as did the incarnation, that the material aspect of the human person is good). In Acts (2:42–47; 4:32–37) the Christian community shares its material goods in common, paralleling the jubilee idea of property redistribution. In Revelation "a new heaven and a new earth" (21:1) are envisioned, where people are nourished and sustained by a "tree of life" (22:2), bringing the Bible full circle to the Genesis paradise: we return to our roots, our material form is affirmed again.

The New Testament teachings that corporeality is an inherent and important part of what it means to be human, and that this physical existence is created and blessed and even experienced (in Jesus) by God, affirm not only human activity in this life but also human care for the earth context of material existence.

The biblical view is anthropocentric but with a qualifier: People are God's images, but are subordinate to God; they have dominion over the earth, but are to care for it in trust from and for God; they are superior to other creatures, but cannot abuse them and must recognize their inherent dignity from God. People as God's images and in their dominion over the earth have status above other creatures; but as people subdue the earth they act as all creatures do to order their environment to meet their individual and species needs, and so have status equal to other creatures in this regard. They are also to share the earth's goods to meet everyone's needs.

The anthropocentric aspect of this perspective might be seen as a culture- and time-based view that might be amended by biblically oriented communities of later times and places with newer understandings of God and creation, depending on the openness of biblical people to those newer understandings. Biblical people must learn to reach out to people in science and of other faiths in order to understand more fully the wholeness of the Spirit's revelation to humanity.

Prophets and Profits—Service to God or Mammon   Throughout the ages religious prophets from different traditions have offered people their understandings of human responsibilities to God. Often these responsibilities included not only worship, but also stewardship: respectful use of the Creator's earth. People are to recognize God's ultimate sovereignty over the cosmos and God's solicitude for the creatures that inhabit the cosmos. Biblical prophets such as Amos, Isaiah, and Hosea, and native prophets of the Americas like Handsome Lake in the east and Smohalla in the west, called for care of the land and compassion for the peoples of the land.

In the biblical tradition, people guided by the Spirit are to work carefully to secure for themselves necessary goods from the earth's bounty. They are to be mindful of others, and, imaging the Creator, compassionately solicitous of others' needs.

In some eras and cultures, the prophets' message became increasingly ignored as people chose individualistic over communal courses of conduct. Lives became self-oriented, leading to a fixation on the acquisition of wealth. As this became more and more possible, because of changing cultural values and laws favoring material accumulation, people had to choose between the older religion-based traditional community concerns about the acquisition and distribution of wealth, and newer attitudes favoring acquisition of wealth solely for self-gratification; they could not be oriented simultaneously in two directions. For those concerned about the teachings of Jesus, the moral course of action should have been clear: "Do not store up for yourselves treasures on earth. . . . You cannot serve God and wealth" (Matt. 6:19, 24). This did not mean that people were not supposed to seek to make a living, or to reap financial benefit in the process; it meant that the acquisition of wealth was not to be their primary goal, and that when wealth was obtained it was to be used for the community as well as for the individual.

On the corporate level, the words of the prophets and Jesus do not mean that companies are not to seek to make a profit: were this the case, business and commerce would not be possible. For the prophets, key questions are:

- How was the profit obtained (for example, at the expense of the worker or the community)?
- How is the profit used (for example, for needed company expenses, including worker salaries and equipment needs, and so forth, and for community benefits)?

Corporate owners, managers, and employees can take legitimate pride in their work and in the benefits that accrue from it, including financial benefits, when these profits flow from responsible use of human skills and natural resources, and flow to responsible provision of individual, company, community, and global well-being. In the spirit of the prophets, then, technology and nature, and corporation and community, are affirmed when these latter conditions prevail.

The Islamic Tradition

In other sacred traditions, respect for the earth is advocated in oral or written teachings. In Islam, for example, the Qur'an instructs the faithful that the signs of God are in creation, in the heavens and on earth, in day and night, rain, life in plants, animals, winds and clouds: "Here indeed are signs, for people that are wise" (S. II. 164). God continues to work in creation, and "to God belongs the dominion of the heavens and the earth" (S. XXIV. 41–46). God provides sustenance to all creatures who need it (S. XXIX. 60–63; S. LV. 10–12). Islamic teaching parallels biblical teaching in its affirmation of the sovereignty of God in creation, and the love of God for creation.

Traditions of Indigenous Peoples

In religious traditions of native peoples of the Western Hemisphere, this attitude of respect toward and affection for the earth and other creatures is pervasive. Among the native peoples of what is now the United States of America such insights abound. The Lakota (Sioux) spiritual leader Luther Standing Bear, for example, declared in *Land of the Spotted Eagle* that "wherever the Lakota went, he was with Mother Earth. No matter where he roamed by day or slept by night, he was safe with her" (pp. 192–93). Standing Bear extends the sense of an earth family cared for by a loving mother to a familial regard for other life forms: "Kinship with all creatures of the earth, sky, and water was a real and active principle. For the animal and bird world there existed a brotherly feeling that kept the Lakota safe among them" (p. 193). Humans and other creatures shared material elements and the presence of the Spirit: "The Lakota could despise no creature, for all were of one blood, made by the same hand, and filled with the essence of the Great Mystery" (p. 193).

The famous Oglala Lakota spiritual leader Black Elk described in *Black Elk Speaks* an earlier era in which his people lived in balance with other creatures and were nourished by Mother Earth: "Once we were happy in our own country and we were seldom hungry, for then the two-leggeds and the four-leggeds lived together like relatives, and there was plenty for them and for us" (p. 8). A Muskogee spiritual leader, Phillip Deere, declared in an interview with this writer that all peoples must understand the traditional indigenous peoples' teachings that humans are related to other creatures on Mother Earth: "We have felt ourselves to be a part of creation: not superiors, not the rulers of creation, but only part of creation. We felt that we destroyed ourselves whenever we destroyed anything within the creation"; and that people are provided for by both Mother Earth and the Spirit: "We must learn to say 'Mother' as well as we say 'Our Father.'"

Other indigenous spiritual and political leaders, from the Americas, Asia, Africa, Australia, and Europe, have spoken similarly of their people's ties to the earth and kinship with all creatures. They have taught that the earth cannot be owned and must be treated with respect, and that all animals who give their lives so that the people might live should be acknowledged with gratitude. The lives of native peoples were often guided by the dreams and visions they had while walking with the Spirit, and while listening to the voices of the earth for insights from the Spirit.

Indigenous peoples' traditions promote respect for the earth—Mother Earth—and love for the Creator and for all creatures, with less anthropocentrism than in scripture-based religions.

Ethics, Technology, and Indigenous Peoples   In their relationships with indigenous peoples, dominant societies often have exhibited four basic types of eco-racism:

1. Exiling native peoples to what the dominating group perceived to be the least desirable lands
2. Diminishing native lands when some desired resource was discovered on them
3. Seeking to dump undesirable byproducts of industrial production on native lands
4. Seeking to locate industrial plants with harmful emissions and effluents on native lands

In all these instances, indigenous peoples lost the use or usefulness, and usually the political, economic, or environmental integrity, of their lands.

In the Western Hemisphere today, from the tundra of Alaska through the mesas of Arizona to the rain forests of Brazil, native peoples still are subjected to the injustices of eco-racism. Government

CONTENT IN CONTEXT 6.3

Saving Sacred Sites

On a spiritual journey up Eagle Mountain one morning, the indigenous elder Soars With Eagles notices some wooden stakes with fluorescent streamers along the western slope. Shortly thereafter he encounters a man in a hard hat who looks at him oddly, then cautiously asks him, "Are you here to protest the construction of the new observatory? I thought the tribal chairman settled everything already. This is an ideal location, atop the mountain: it's almost like being close to God." Soars With Eagles had not heard about this construction. "You say that Chairman MacDougal approved it? How could he possibly sell this site?" The worker looks nervous, so Soars With Eagles bids him goodbye and returns back down the mountain to his village. He visits other elders to see what recourse they might have to convince their people (many of whom have forsaken traditional beliefs, and have stopped participating in traditional religious practices) to prevent their ancient prayer site from being desecrated by construction, polluted with vehicle fumes and sewer effluents, and overrun by scientists and tourists. "The tribal chairman has sold us out for a few dollars and the promise of new jobs and government aid. What should we do now?" Dancing Wolf, another spiritual leader, agreed that they must act soon or else all of their sacred sites might be destroyed in the name of "progress," which "usually meant that more lands would go out of our hands" and be used by and for the commercial and industrial projects of outsiders. "Let us smoke the sacred pipe and pray for guidance."

In what ways might the conflict in cultures described be avoided? Should the sacred site be protected from alteration for scientific progress? If your response is "no," would you maintain your position if the sacred site were a cave in Bethlehem, a mosque in Mecca, or a temple in Jerusalem?

and corporate strip-mining, deforestation, plant siting, waste disposal, and displacement operations threaten the lives, livelihoods, and lands of indigenous peoples. These operations promote physical and cultural genocide, and violate treaties, national laws, and international human rights covenants.

Ethical technological development—which would also be environmentally viable and economically sustainable—must take into consideration the rights and needs of indigenous peoples. These peoples of distinct races, languages, and cultures should participate in decisionmaking processes that concern their lands. They should have adequately informed traditional representatives to present their views, not spokespersons chosen by government (such "representatives" often are those who have become acculturated to the values and practices of the dominant society). Indigenous peoples' interests would best be served if they had some measure of autonomy over their lands. Indigenous peoples' spirituality, so complementary to the religious beliefs of other traditions, can help to guide responsible technological innovation by linking human creativity and productivity with an ethical concern for the rights of all creatures and for the well-being of the earth.

GENDER EQUITY IN ECONOMIC AND ENVIRONMENTAL ISSUES

Throughout the ages, religious traditions often have had a patriarchal structure and ideology. Although this bias has been particularly true of scripture-based religions, it has also been present in some faiths based on oral tradition. In some cultures, the religious patriarchal bias infused political and economic structures. In Western societies, for example, women were denied the right to vote or to run for public office, and were denied equitable remuneration for their work because they were expected to be the wives of working men and the mothers of their children.

In the latter part of the nineteenth century and into the twentieth century, the bias against women and the oppression it fostered have been somewhat ameliorated in some societies and within some religions or denominations of religions. On the ideological level and sometimes on the practical level, gender equity is being fostered in economic issues and in environmental issues. Women's struggles

CONTENT IN CONTEXT 6.4

Equal Pay?

Clara Darrowe worked at the Soo, Cashe and Phee law firm for ten years, working her way up from assisting other lawyers to representing top clients. Expecting a promotion for her work, Clara was shocked when Ken Soo announced that the next senior partner would be Bill J. Bryann, a young attorney who had defended the Cosmic Rocket Company against a multibillion-dollar suit by an independent engineer, Alicia Einstine, who claimed that Cosmic's latest magnetic thrust engine had been adapted from a patent that she had filed. Bryann's innovative argument was that telepathic transfers had allowed Einstine inadvertently to appropriate ideas from Cosmic's engineers, who were working on a similar invention (in subsequent court cases, this would be cited as the "telepath defense"). Einstine's attorney argued: "If that were the case, why had Cosmic not gotten a patent? And if both engineers had the idea simultaneously, in a concurrent burst of creative resolution of the same propulsion problem [this would later be cited as the "concurrent congruent creativity" factor], then Einstine should be entitled to receive at least half of the royalties for the product, especially since she had filed the patent." The judge ruled that what was key for him was the idea of "putting your money where your mouth is," and since Einstine had not produced a marketable product, while Cosmic had produced a working model, she had no royalty rights. As Clara thought about the case, she realized that she and Alicia were the recipients of similar treatment from Soo, Cashe and Phee: each woman had worked hard, but both had been denied the professional recognition and financial reward they deserved from their labors.

Is Clara Darrowe's assessment accurate? Can you cite any examples of gender discrimination in the workplace? How might gender discrimination be eliminated?

have enabled them to take a greater role in guiding society to an increased consideration and elimination of economic inequity and environmental degradation.

In the economic arena, women have rightly obtained greater freedom to become entrepreneurs, have achieved recognition for their abilities in technological innovation, and have begun to acquire greater equity in compensation for equivalent types of work and greater opportunities for advancement within their company.

In the area of environmental concerns, the special needs of women in certain types of industrial processes has begun to be recognized, sometimes in negative ways and sometimes in positive ways. Negatively, some industrial corporations, upon becoming aware of the dangers some chemicals pose to fetuses, have required women workers, as a condition of employment, to choose sterility or lower-paying safer jobs to avoid exposing the life within them to toxins that might cause deformity or death, or to hold the company exempt from responsibility for fetal devastation resulting from exposure to harmful chemicals. Positively, other companies have altered production practices so that the health of women and their potential offspring, and the integrity of their reproductive system, would not be impaired by industrial production.

As humanity heads toward a new century, the gains made by women in economic and environmental areas of concern should be extended so that eco-sexism might be eliminated from society, to the benefit not only of women, but of their families, society in general, and the earth as well. Religious thought and practice can greatly assist the transformation of society to achieve respect for women's rights and needs.

INTERPRETATIONS OF RELIGIOUS TRADITIONS

A problem arises with the study of religious traditions to interpret human relationships with the earth when readers isolate the written texts or oral traditions from their historical contexts and from the larger corpus of which they are a part.

When studying sacred written texts, people sometimes use *eisegesis*, reading their own ideas into a text to make it conform to their existing viewpoint, rather than *exegesis*, reading a text critically to discern its meaning. Eisegesis sometimes includes a failure to distinguish between a writer's use of history and a writer's use of legend or allegory or story to teach a religious insight. In addition, people sometimes fail to distinguish between a writer's culturally related perception of what the Spirit requires or hopes for from humanity, and a writer's inspiration from the Spirit: in the former instance, a writing might prove useful for a time in a given location; in the latter case, the message is more universal and enduring.

In the effort to analyze a sacred written text such as the Bible (exegesis of a passage) and to understand in the process the influence upon that text of ideas and events in a historical era (cultural perceptions incorporated into the passage), the interpreter must try to discern if the teaching in a particular writing is (*a*) valid for all times and places; (*b*) valid for a particular time and place, appropriate to the people of that particular type of historical situation (in terms of the moment of its adoption and later equivalent moments); or (*c*) advocated due to a misunderstanding of the divine will, a misperception of historical or scientific reality, or a misuse (intentional or unintentional) of religious authority.

In the Hebrew Scriptures/Old Testament, for example, we read in Leviticus 11:7 of the prohibition against eating pork. Contemporary Jewish scholars (and Jewish laity) argue over whether that biblical prohibition continues to be valid. The text might seem to indicate an ongoing prohibition, a law that is valid for all times; but the counterargument is that the law was necessary and useful for a desert people lacking the refrigeration, understandings of good health practices, and government regulations we have today. For those who support the former view, changing historical circumstances have no impact on the biblical law: pork is still prohibited. Those who disagree point out that while the law was necessary to protect the health and lives of earlier generations, it is no longer needed today in areas where meat and meat production and processing facilities are inspected, where meat is refrigerated or frozen, and where consumers know how to cook pork properly. The two positions cannot be reconciled: a devout Jew is either prohibited from eating pork or permitted to eat it. The course followed depends on whether the particular law is believed to endure perpetually or change historically, whether it was meant for particular circumstances or must be observed in all circumstances.

(In the Christian Scriptures/New Testament, the issue of pork consumption and, by extension, of whether or not some religious laws are universally and perpetually valid is resolved, in part, in Peter's vision in Acts 9:10–16: Peter sees a tablecloth unfold from the sky, laden with a variety of animals, some of which in the form of food were prohibited to him as a devout Jew. A voice tells Peter to "kill and eat." Peter protests that he cannot eat "unclean"—religiously prohibited—food. The voice tells Peter, "What God has made clean, you must not call profane." Peter comes to understand that neither he nor other Christians need to observe some older religious laws.)

The Bible (and every other sacred text) is a religious book, not a science book or a history book. When people try to make it a science book, they focus not on the religious teaching but on the details through which that teaching is conveyed. For example, a biblical writer seeing the sun travel across the sky might assume that it revolves around the earth. Subsequent generations might observe that apparent passage of the sun with the naked eye, read their Bible, and then condemn the scientist—Galileo—who, having seen something different with his telescope, teaches that the earth revolves around the sun. Those who try to maintain that a detail is a revelation in such circumstances do violence not only to the scientist and to the truth of objective reality, but to the sacred text itself; they advocate what the text does not teach, causing confusion among those who learn a demonstrable reality from science teachers but hear a distorted interpretation of reality from religion teachers.

Exegesis linked to religious tradition will illustrate the problem of understanding the extent to which a teaching is valid for all times or for only a particular time, or whether it has some insight valid for all times with details that might be change historically. Take the issue of theft. There might seem to be no problem. Surely theft is wrong and illegal: does the Bible not say, "You shall not steal," and do our laws not reinforce that command? But what about the case of someone in dire need: hungry and unemployed and penniless, for example?

In contemporary Western societies, theft seems simple to define: the taking of another's property for one's own use, "another's property" usually being defined as legal ownership. The simple definitions just cited become less certain in the light of certain biblical and theological understandings.

In the Bible, some property has a community primacy: land, and the goods of the earth and products made from them. So, in the Old Testament/Hebrew Scriptures, a hungry person is allowed to enter a neighbor's vineyards or fields to satisfy his or her immediate hunger with grapes or grain (Deut. 23:25–26). In the New Testament, we see Jesus' disciples doing just that as they walk through a grain field and pluck heads of grain to satisfy their hunger (Matt. 12:1). As time went on, and Christianity became the dominant religion in Europe, this right to satisfy one's needs that took precedence over private property rights was forgotten.

In the Middle Ages, however, the great philosopher and theologian Thomas Aquinas (1225–1274) reminded people that needs take precedence over property. In a nonconfrontational version of the philosophy and tactics of England's legendary Robin Hood, Aquinas declared unequivocally in his *Summa Theologica* that "in cases of need all things are common property" (II, II, Q. 66, Art. 7). In Aquinas's thought, those who have an abundance of goods are bound by charity to help those in need. If they are not fulfilling that responsibility, then someone who has an urgent need may acquire necessities on their own, even to the extent of taking another's unneeded "property." For Aquinas, "it is not theft, properly speaking, to take secretly and use another's property in extreme need: because that which he takes for the support of his life becomes his own property by reason of that need" (ibid.). Moreover, "in case of like need a man may also take secretly another's property in order to succor his neighbor in need" (ibid.). Aquinas believed that "according to the natural law all things are common property" (II, II, Q. 66, Art. 2). He taught that private property was devised by human reason as an addition to the natural law. Since the natural law comes from God, laws protecting private property are secondary to the requirements of the natural law that people's needs be met by the goods of the earth. Need, therefore, takes precedence over property.

The teaching of Aquinas did not disappear as time went on. In *The Catechism of the Catholic Church*, issued by the Vatican in 1994, the Catholic Church teaches in the section on the commandment prohibiting stealing that "the goods of creation are destined for the whole human race," and that this universal destination of goods is primordial over private property. The church then states forcefully that while "the seventh commandment forbids theft, that is, usurping another's property against the reasonable will of the owner," theft does not occur in the case where there is an "obvious and urgent necessity when the only way to provide for immediate, essential needs (food, shelter, clothing . . .) is to put at one's disposal and use the property of others" (par. 2408).

Recent church teaching on the issue of theft, then, reiterates the scriptural teachings revealed by exegesis. Eisegesis, by contrast, might reinforce property relations that justify the position and possessions of a particular social class, or seek to justify an unjust economic structure in a particular nation.

Over time, eisegesis-generated interpretations and culturally oriented perceptions become increasingly irrelevant. They contradict scientific knowledge, literary analyses, and religious insights. Those who continue to hold to such interpretations and perceptions tend to become more individualistic, rigid, and even fanatical as their ideas become more marginal to societal beliefs.

By contrast, those who utilize exegesis, and who remain open to the work of the Spirit in the world through the ages and across cultures, become more communal or community-oriented, in both a social and a scientific sense: they find points of identity in each cultural context and age, and areas of relationship with new understandings and fields of study. Spirit and science are bridged in such people and in the organizations that share their perception and perspective.

The major consequence of the reflections and actions of the exegesis-based, Spirit-seeking person is that faith and science are woven together in human relationships with the earth. The ethics that result, on both the theoretical and the practical level, might serve to guide scientists and spiritual people (and those who are both). They are based on a holistic understanding of the cosmos and of humanity's place in it, and consequently can lead to a responsible use of the machine (technology) and a respectful sojourn in the garden (the earth).

THE MACHINE AND THE GARDEN

One of the enduring myths in Western (and other) traditions is that of the garden. People postulate the existence of an ideal garden in the past in which all creatures lived in harmony with each other and

CONTENT IN CONTEXT 6.5

The True Prophet

Jeremiah Josephson was walking along Main Street one Sunday afternoon when he saw a small crowd gathered around two men on wooden crates who seemed to be arguing with each other. Curious, Jeremiah approached the group to listen in on the debate.

"Eliminate technology! It is destroying the earth! Return to a simple life!" shouted one of the men.

"Embrace technology! It is enhancing our lives! Don't live in a Stone Age!" screamed the other.

"The earth is in danger! We're consuming too much! We're too dependent on things!"

"We're living longer! We're traveling faster! We can conquer illness! We can fly to the moon!"

"God is angry! We'll be punished like Adam for eating the fruit of knowledge, and have our machines destroyed like the Tower of Babel!"

"We're in God's image, called to be creators and inventors! Don't listen to him! Remember what his kind did to Galileo!"

Jeremiah listened to the debate, thinking that each side seemed to have some valid ideas, but neither seemed to be willing to reach a compromise with the other's position. Slowly and reflectively walking away, he considered that he drove a car to work and did his reports and budgets on a computer; but he also liked to hike on the mountain and bike through the park to get away from things.

Would you sell your car, and walk or ride a bike or bus everywhere? Could you endure a concrete world with only videos of parks instead of real parks? What would be gained and lost if either side were to dominate society? Do people need parks and computers? How can the machine (technology) enhance rather than endanger the garden?

their Creator until something shattered the idyllic setting. In more modern times, the myth of a paradisal garden has been coupled with the sobering thought that humanity brought a machine (science, technology, industrial ventures) into the garden which disrupted the natural flow: therefore people must choose between the ideal garden and the real world, since technology and paradise seem in perpetual conflict, even though technology was introduced into the garden to recreate its original wonders. A more optimistic view would be that the garden and the machine can peacefully coexist—if the former is treated with respect, and the latter is used with care.

The spiritual insights presented above can help to overcome both religious misinterpretation of sacred traditions and religious antagonism toward the physical sciences, social sciences, and technology. Use of these insights will enhance discussion of the faith–science dialogue and the machine–garden debate.

SCIENTIFIC INSIGHTS FOR AFFIRMING TECHNOLOGY AND NATURE

What has been stated about religion, about its possibilities of individualism, communalism, and fanaticism, might also be stated about science. Science might focus solely on individual research and private benefit; on teamwork and community benefit; on unrestricted research and closedmindedness; or on any of the preceding with also some openness to possibilities beyond the quantification and understanding of a particular scientist or scientific school of a given historical moment.

Science, and the technological innovations that result from scientific creativity, can provide for people and planet a brighter future if research and development respect social needs and environmental exigencies.

Religion and science can be mutually corrective. Science can correct religious efforts to make the "ordinary" something "extraordinary," that is, to create supernatural explanations for occurrences that result from the interaction of physical, chemical or biological laws and forces. Religion, in turn, can correct scientific efforts to convert the "extraordinary" into the "ordinary," to declare that an unusual occurrence must either be the result of some as yet undiscovered law or force or else could not possibly have occurred because it does not fit into current or projected scientific understandings. The perspective of religion should be to explore to the greatest extent possible scientific explanations for unusual events, and to accept verifiable explanations; the perspective of science should be to accept not only limitations on human knowledge in the present stage of technological development, but even the possibility that in the future there will still be limitations, because some realities cannot be quantified by technologies no matter how sophisticated: they are simply of a different order of things that is not scientifically measurable. People in both fields should have a certain sense of humility and respect before the wonders of the cosmos and before the wonders of the Spirit.

Scientific analysis can correct religion's sometimes romantic evaluation of the world; scientific knowledge can correct uninformed religious declarations about human nature and geologic time. Religious compassion can correct dispassionate scientific calculation about societies and resources; religious moral sensibilities can correct uninformed scientific presuppositions about human nature and conduct and the consequences of human activity. In short, religion needs science's analytical perspective; science needs religion's moral passion.

Religion and science must both move out from under the shadow of the trial of Galileo, in which religious insecurity, ignorance, and arrogance used power to negate the truth of science with biblical myths. The religionist and the scientist must jointly excise the Galileo syndrome: the fear on the one hand that science would undermine faith; the fear on the other hand that religion would persecute legitimate science once again. When science is respected, and spirituality is revered, Spirit and science become the joint foundations of a new ethic that links technology, community, corporation, and environment.

ENVIRONMENTAL IMPACTS OF POPULATION, PRODUCTION, AND POLLUTION

One area in which scientific investigation has proven helpful in analyzing human interaction with the rest of creation has been in the relationship between population growth, expanded industrial production to meet the needs and wants of the increased population, and pollution resulting from both population and production.

The earth is a limited land area: its carrying capacity, although not known yet, is assuredly finite. Increased inhabitation brings increased need for land for housing, agriculture, and employment; increased demands for water for home, farm, and industry; and increased stress on air needed by people, crops, livestock, and other life forms. Scientific research in the area of population growth has revealed the potential for a catastrophe if that growth remains unchecked, and has offered several courses of action for maintaining or decreasing populations nationally or globally. Since space travel and settlement are still infeasible, and wars and plagues are undesirable, birth control methods appropriate to human and planetary needs, and responsive to religious concerns, are being developed by science to offset the harms to creation occasioned by people's lack of reproductive responsibility (sexual self-restraint) and lack of resource responsibility (consumption self-restraint).

Increased inhabitation also brings increased pollution, particularly in Western industrialized countries but also in poorer nations seeing to emulate the levels of economic development and "standard of living" of the richer countries. In the past, research and development efforts often concentrated on creating and marketing new products for consumption, usually with inadequate investigation of the consequences of those products, of the impacts of their production processes, or of the byproducts of their production. Communities are left today with massive cleanup requirements as a result of those activities. Cleanup activities have catalyzed, in turn, new scientific research into both restoring the environment and eliminating the need, through creative planning, of future cleanups.

In this field, then, science and technology are helping to restore the earth and to provide ways by which the earth's carrying capacity will not be exceeded. As these efforts prove successful, population will be stabilized, production will be planned carefully, and pollution will be diminished.

NEW TECHNOLOGIES: NEW PROBLEMS OR NEW PROMISE?

In the past half-century a wealth of new technologies has been developed and marketed to meet human needs and wants (the latter sometimes skillfully created through advertising). In the areas of communications, medicine, and warfare, to cite a few, these technologies have altered forever the relations of people with each other and with the earth.

New technologies have brought problems: weapons development and genetic manipulation, among others, have rendered human survival more tenuous. People can more easily be killed by new types of explosives, some contained in weapons launched from thousands of miles away; and the environment can more easily be devastated by new life forms, developed in laboratories, which have no natural deterrents.

New technologies have also held out the promise of benefit for people and planet: oil-eating enzymes developed in one technological field can offset the pollution developed because of failures in another; biotechnology can be the key to understanding the generation of disease and to intervention to confront or eliminate it; and telecommunications can unite peoples around the world to promote understanding and mutual respect. Religious insights and ethical considerations can help to promote the promise and diminish the danger of new technologies.

SPIRITUAL AND SCIENTIFIC INSIGHTS: CREATING RESPONSIVE AND RESPONSIBLE TECHNOLOGY

When spiritual inspiration and scientific insight are perceived to be in conflict, or mutually exclusive, human advancement and environmental well-being are placed in jeopardy. By contrast, when the spiritual values and scientific ideas discussed above are related objectively to each other, they can help to create an atmosphere for the responsive and responsible development and use of technology, a type of technology that is responsive to human and environmental needs and responsible in the way it goes about meeting those needs. Such a technology will be appropriate to the materials and task at hand and to all aspects of its biome, and thereby sustainable. (Biomes are areas in different parts of the world that have similar flora and fauna, and similar patterns in soils, climate, and hydrology. There are fourteen basic biomes on the earth.)

New conversations have taken place in the last few years between religious and scientific professionals, and some of the old rivalries and mistrust have been diminished. Much remains to be done to go beyond discussion to development, from insight to implementation: to joint efforts to create a more just society and a healthier planet. As these efforts take place, appropriate technologies will be developed that will responsibly help to meet human needs and the requirements of the planetary ecosystem.

When spiritual values, scientific insights, and social concern are integrated together in planning efforts for social betterment and in concrete projects that demonstrate compassion for people and care for the earth, ethics and technology jointly will stimulate truly progressive community development.

ETHICAL PRINCIPLES

In the spiritual tradition, human understandings of the Spirit and creation might be expressed as follows:

1. The cosmos is in the Spirit, its Creator.
2. The earth is entrusted by the Spirit to human care.
3. The earth should be treated respectfully, and its fruits should be shared equitably through the ages.

In a humanist tradition, these principles might be expressed in a way that does not acknowledge the Spirit but affirms the human obligation to be respectful toward the cosmos:

1. The cosmos should be respected.
2. The earth's well-being is assisted by human care.
3. The earth should be treated respectfully, and its fruits should be shared equitably through the ages.

CHAPTER SEVEN

Tradition, Technology, and Transformation

As we look around us, it is evident that the ethical principles we have considered are not widely put into practice in the world. We observe problems of unemployment and underemployment, poverty, pollution of air, land, and water, conflicts between local communities and among international communities, and conflicts between companies and communities. The next task before us in the management and utilization of technology is to find ways to transform our local and global communities so that technology is wisely and equitably used, and so that the future reality in which our descendants will live will more closely conform to our best dreams for the earth and for future generations.

The transformation to a better world can take place if technological innovations take into consideration ethical traditions. When innovators are conscious of and respond to the needs of individuals, corporations, communities, and the earth as a whole, their innovations can lead to individual, corporate, community, and planetary well-being:

$$\text{tradition} + \text{technology} = \text{transformation}$$

The union of ethics and technology can transform the world. The appropriation of social ethics by innovators, entrepreneurs, citizens, and communities can lead to the development of appropriate technologies and to the creation of good jobs that help to provide for people and to preserve the planet.

TRADITION

In the humanist and religious traditions that are operative in the present era, concern for the careful development of the earth to meet human needs and concern for the long-term viability of the earth and all its organisms are linked together as people—in their role as citizens or in their role as entrepreneurs or employees—seek to promote both employment and environment to sustain present and future generations. In the process of promoting positive change, people must reject those aspects of tradition that are harmful and preserve those that are helpful.

This rejection–preservation scenario is especially important today in considerations of the relationship between ethics and technology within the broader context of the relationship between religion and science. At times in the past when religion or science strayed from its respective field of competence, the two came into conflict. At other times, they found fruitful areas of collaboration. The earth today is experiencing an ecological crisis, and humanity is experiencing a societal crisis, and the two are related: some individuals and groups have devastated natural systems and dominated social structures for personal, corporate, or national benefit. As we look for the causes of this condition and seek ways to restore balance to nature and societies, we might blame religion and science for their respective contributions to the present problem and, at the same time, bless them for the perspectives they offer that contribute to its future resolution. Religious thought and practice have both blessed the earth and cursed it, have advocated both involvement in this world and escape from it. Scientific thought and practice have both helped the earth and harmed it, have brought about both its exploitation and its conservation.

When the worst perspectives of either religion or science permeate human activity, a rejection of the rational or spiritual occurs, and humanity and the earth both suffer as a consequence. When the best perspectives of both religion and science permeate human activity, a synthesis of the spiritual and rational becomes possible, and human progress on several levels, within a context of relationship with the earth community, may be both envisioned and concretized. Religious imagination and scientific imagination working together unleash creative powers in a synergy unavailable to either by itself, and this creative force re-presents the creative cosmic act, freeing people once again to be images of the Creator Spirit.

The quest for meaning in human existence can lead people along diverse paths. Some seek answers to profound questions solely in science, and use reason to explore concrete, demonstrable reality. Others seek solace solely in religious faith or in the search for spiritual experiences, and either use reason to probe religious doctrines or set reason aside in efforts at direct prayerful engagement with the Spirit. Others explore the intricacies of philosophical inquiry, and compare or construct intellectual speculations about ultimate meaning. In the end, no single way might be complete in itself for someone who seeks the meaning of life: the material world seems finite and perishable; the spiritual world seems nebulous and unverifiable; and the intellectual world seems abstract and theoretical.

Human reason can be informed by the insights of any and all of these paths, such that imagination and creativity are unleashed for a new synthesis of the scientific, the social, and the spiritual. When individuals or groups do not allow themselves to be constrained by one narrow field of inquiry, their capacity for understanding variant aspects of existence and for engagement with distinct types of realities is enhanced. Spiritual reality is found when reason is unbound. At that moment, spiritual tradition and scientific tradition become aspects of one reality.

TECHNOLOGY

Technology has an important role to play in creating a sustainable planetary future. Human ingenuity has invented technological innovations that enhance human existence. There have been some technological developments that have been—and are—harmful to people and planet, and efforts are under way to eliminate some of them, or at least to ameliorate their harmful impacts. When technological developments are guided by ethical insights, the earth and society both benefit.

TECHNOLOGY AGAINST TRADITION

Some scientists today still reject the relevance of ethics—especially religious ethics—for technological innovation. Their reasons for this rejection include the following:

CONTENT IN CONTEXT 7.1

Origins of Species and Spirituality

Dr. Brenda Washington is a biologist, the faculty adviser for the African American Studies Institute at her university, and a soprano in her church's gospel choir. As she prepares for class one evening, she reflects on how the scientific, ethnic, and religious aspects of her life weave together. Her lecture on the following day will focus on the theory that all human life had its origins with an African woman designated "Eve" by the scientists who analyzed the human genetic code, compared their findings with human genetic traits around the world, and decided that humanity originated in Africa. Dr. Washington recalls the biblical creation story from which the name Eve was taken and wonders about the relationship between her work as a scientist and her religious faith. "Are people and the cosmos of which we all are a part created, or is everything just combinations and recombinations of eternal atoms?" She hopes that there is a higher being who might provide a deeper meaning to transitory human existence, but realizes that her choices are to have faith in God, or to continue to speculate about God, or to deny spiritual reality, since science cannot prove a Spirit's presence in—or even absence from—the universe. "If a Spirit does exist, how does that Spirit view my work: as an intrusion into the divine order or as co-creative activity? Maybe it depends on the type of work I'm doing, on the purposes for which I'm doing it, and on the respect I have for the complexity of the universe about which I speculate, parts of which I manipulate."

What relationship might there be between Dr. Washington's work as a scientist and her religious faith? When might the Spirit, if one exists, regard her work as an intrusion into the divine realm, and when as co-creative activity related to divine purpose? Do you believe that science can explain or will be able to explain every mode and relationship of existence, and every type of reality?

- Scientists claim to be "value neutral."
- Scientists want no interference with ethically questionable research and development.
- Scientists have legitimate fears, based on very real historical incidents, that people professing ethical concerns or religious creeds will unfairly and uninformedly interfere with legitimate scientific endeavors.

In the first instance, the claim to be "neutral" when innovations are potentially harmful is invalid. In the second instance, the need for ethics obviously is apparent to protect people and planet. In the third instance, common ground must be found for scientists and people of faith in positions of power to respect each other's areas of competence and authority, and to work together to meet ethical, economic, employment, and environmental needs.

TECHNOLOGY AND TRADITION

Technology presents some of the highest achievements of the rational mind. It is similar to literature, theology, art, and philosophy in that engagement with it unleashes the human imagination and human creativity to probe the human condition, offer insights into it, and suggest ways to improve it. The innovative process in technology, coupled with ethical insights relative to justice and concern for creation, can lead society to material well-being, social stability, and environmental sustainability. Tradition and technology, when linked together, can effect social transformation.

TRANSFORMATION AS CONTEXT, CONFLICT, AND CONVERSION

Every social context carries the seeds of its own renewal or transformation. Through education, experience, creativity, technological innovations, and spiritual insights, individuals and groups are able to analyze what is and project what should be or could be. They can compare their vision of what the world might be like to the reality of the world as it is, and propose concrete historical projects to change what is into what should be: to make the new local, national, or global reality of the future congruent with their present vision.

If the current historical context is regarded as a topia (place), and

CONTENT IN CONTEXT 7.2

Galileo or Nietzsche?

Dr. Carla Darwen is a paleontologist exploring an exciting new discovery of human antecedents in Kenya. She happily catalogs the skeletal remains of an ancient forager, thinking, "I've uncovered one more proof of the scientific errors in the Bible regarding human origins." Dr. Darwen takes pride in disproving religious superstition, and in affirming scientific objectivity. She is somewhat uneasy that some Christian groups in her native land might object to her findings and question her credibility as a scientist, but she reassures herself that these groups are losing power as the Bible's "science" becomes more and more incredible. The rest of its teachings are outmoded too, she thinks with a smile: "Religion is outdated and God is nonexistent and irrelevant: we don't need a god to rule us. Science can meet human needs and solve human problems, including problems created by science and technology." Dr. Darwen smiles again when she remembers some of the religious artifacts she and her colleagues discovered among some ancient remains. "Those folks had an excuse for their superstitions: they didn't have the tools and insights we have today to unlock the secrets of the universe. We don't need any spiritual or extraterrestrial beings to enlighten us and to give meaning to our lives." Sensing a presence, Dr. Darwen turns around. An old native woman in colorful dress standing behind her, leaning on a cane, and staring at the bones in her hands, asks: "Why are you digging up our grandmothers' bones? Have you no respect for the spirits of the dead?"

Is the possibility of religious trials for scientists real today—if not in court, in media attacks? Can Dr. Darwen validly extend her conclusion that the Bible is erroneous in scientific areas to assert that the Bible—or any religious tradition—is in error in affirming spiritual realities and ethical insights? What respect should scientists have for native peoples' burial sites and traditional beliefs?

the hoped-for context is a utopia (no place: not now existing), a conflict might take place—on an intellectual or a societal level—between reality and vision. This conflict might be represented as follows:

$$topia_1 <\text{—}> utopia_1 ====> topia_2 <\text{—}> utopia_2 ...... ===> topia_n <\text{———}> utopia_n$$

That is, the conflict between what is and what might be leads to a new context, which in turn is in conflict with new visions, and so on.

CONCEPTS AND CONTEXTS OF UTOPIA

Three points are important in an analysis of utopia. First, there is a distinction between an absolute utopia and a relative utopia. The absolute utopia is the fully transformed society of the future, or the fully humanized resolution of a particular social issue in the future. The relative utopia is a partial realization of the transformed society, or the partially realized resolution of a particular issue. Relative utopias are steps toward achievement of absolute utopias.

Second, there can be either a positive or negative connotation of the term "utopia": positively, it means a context which is not yet existent, but could exist; negatively, it means a context which is not existent and can never exist.

Third, of key importance to a society (and to the world) is who defines what is "utopian" in a negative sense. Those who benefit from the status quo seek to dismiss as impossible dreams the realistic utopian visions of social reformers. Social visionaries and reformers must hold on to their positive utopias when these are derided or denied. They must plant the seed, even if, as when planting an acorn, they might not benefit from it or perhaps even see it come to maturity.

In the area of technology, numerous "givens" of today would have been termed utopian in the negative sense just a decade or a century ago. Had creative scientists listened to those who accused them of being dreamers, then satellites, space shuttles, submarines, computers, and modern medicines would still be the stuff of science fiction novels. Similarly, in the area of social justice, ideas about minority rights, women's rights, and environmental responsibility were once disparaged as contrary and dangerous notions (and still are today, by

CONTENT IN CONTEXT 7.3

Mountain Visions

It is a beautiful day for a hike. Dr. Carla Darwen, back from her anthropological research in Kenya, had telephoned her friend Dr. Brenda Washington to suggest a climb to the mountaintop. The two spend three hours on the hike. They weave among the trees, watch an eagle circle above, and glimpse the white tails of two deer bounding away at their approach. At the summit, they are rewarded with a spectacular view of the city to the east and the forest to the west, all topped by a blue sky decorated with puffy clouds. "It's beautiful!" exults Brenda. "Wouldn't you like it to always be like this?" responds Carla. The two friends hear footsteps on the rocks, and turn to see a native elder wearing black boots, blue jeans, and a flowery green shirt with colorful ribbons sewn across the front and back. His long gray hair, tied in a ponytail, swings from side to side as he climbs over the last boulder.

"Hi, I'm Carla and this is Brenda."

"I am called Soars With Eagles. Usually, I climb a different mountain."

As Carla, Brenda, and Soars With Eagles look below, they see the bustling city with its industrial plant bordering an American Indian reservation nestled against the western forest, part-way between the city and the trees. The plant's smoke drifts over the city; the reservation is dotted with abandoned cars, signs of pervasive poverty. The forest stretches from the end of human encroachments toward the distant horizon.

Complete the story. What conversation might the three have? How might they analyze the present, and what might they envision for the future? What values might they share, and what aspects of their traditions might be complementary? In which types of projects might they jointly become engaged in order to concretize their shared vision? What is your own vision of a better relationship between spirit, science, and nature? With whom might you share your vision, and with whom might you work to make your vision a reality?

some!), but have begun to be established in intellectual and geographic places. As humanity looks toward its next century and millennium and beyond, it is important that utopian visions in technology and in social ethics continue to challenge individuals and societies and seek to draw them forward.

Context, Conflict, and Conversion

The context and conflicts of every present can lead, through conversion, to a better future: for humanity, for other life forms, and for the earth. Conversion, in an individualistic and materialistic social setting, means that people change their way of thinking and their manner of acting from competition to cooperation, and from selfishness to social concern. When that happens, collaborative community-based and community-oriented projects result—to the benefit of people and planet.

Technological innovation can lead to a historical transformation that includes global betterment and the development of new persons in new societies. It can begin as a utopian idea and end as the foundation of a new and better topia. Ethics and technology, working together, can promote not just immediate survival but also ongoing economic viability and economic equitability for human societies, and environmental sustainability for the earth as a whole. Ethics and technology can represent the best of human aspirations and human creativity, and jointly can help to restore respectful relationships among peoples and between people and the earth.

CREATING A NEW FUTURE

The future is the parent of the present: the vision we have of what we want the world to be like will guide our activities in the world as it is, so that our social projects are continuing efforts to alter our topia to make it ever more congruent with our utopia. The present is the parent of the future: what we imaginatively conceive and labor for today will be born as the world of tomorrow.

When we use our imagination and work to concretize what we see as possible, we create our world; when we neglect our imaginative powers, or reject (out of fear of change, or fear of controversy, or fear of commitment) what they offer as possibilities, we are created by our world.

Without a vision, people, like a pond without new rain, will stagnate and perish. With a vision, and through projects to realize it, people, like a river renewed by underground springs, will flow fresh and free.

Tradition and technology can be the springs that unite to create social transformation. At the intersection of ethics and innovation, imagination is the opening that allows the springs to flow into the river of humanity. The spring waters become a renewing force in the river as it interacts with the land, the air, and the sun, and all the creatures in and around it.

In historical time, humanity approaches a new millennium. People have the opportunity, with the insights of tradition and technology, to envision for the next millennium an era of justice, peace, and ecological harmony, a time in which economic viability, political stability, and environmental sustainability will be inseparably linked. For some, human ingenuity, compassion, and commitment will be sufficient to catalyze the coming of that better future; for others, spirituality will be interwoven with such human concern and creative activity in order that a new world—a transformed earth—will be born. In either case and course of action, humanity has the opportunity to overcome and avoid the social injustices and environmental harms of the past, and to create a better future for the planet and for the people and other life forms of the ages yet to come. Humanity, in its material aspect confined by time and poised in time, can build upon the best of its historical traditions—scientific and spiritual—to renew the structures and stage of its earthly existence.

Each generation has its own moments of vision, its times to see beyond the visions and practices of past generations and to formulate and initiate social projects that will effect a better world. Generations overlap; every moment can be a moment of vision. It is our time now to begin the transformation of our ideologies and of our activities so that the reality of future generations will be congruent with the better world we envision today.

Now is our moment to envision and to create a new earth.

APPENDIX A

Case Studies

The case studies included here offer representative examples of organizations whose philosophy and activities embody the ideas presented in selected chapters.

CHAPTER 3: MONDRAGÓN

In the Basque country of Spain during the Franco dictatorship a Catholic priest, José María Arizmendiarrieta Madariaga (1915–76), planted the seeds of a new type of worker cooperative in the little community of Mondragón in the province of Guipúzcoa. Arizmendiarrieta established the Escuela Profesional in October 1943; five of the graduates of this professional school would found the first Mondragón cooperative, under his direction, in 1956. This cooperative, Ulgor, manufactured oil heaters and ovens. By 1985, Mondragón included 172 cooperatives: 94 industrial, 9 agricultural, 1 consumer, 44 teaching, 17 housing, and 7 service enterprises whose individual membership ranged from four to fourteen hundred. Since its beginning in 1956, Mondragón has had an almost 100 percent survival rate among its cooperatives. After little more than three decades, its membership has grown from the original five incorporators to more than 25,000 worker associates.

At the core of Mondragón is its Caja Laboral Popular, a cooperative bank (begun in 1960 with two employees) that guides all the cooperatives. Its governing board consists of twelve members: four elected from the bank cooperative itself, and eight elected from the other member cooperatives.

Basic Principles of the Mondragón Cooperative

In 1987 the First Congress of the Mondragón Cooperative Group proclaimed ten "Basic Principles of the Mondragón Cooperative Experience":

1. Open membership. Mondragón declared itself open to all men

and women who accept the ten basic principles and demonstrate their professional competence for open positions.

2. Democratic organization. Mondragón proclaimed the basic equality of all of its worker associates, as demonstrated in their acceptance of the cooperative's democratic organization. This organization included a sovereign general assembly (composed of all worker associates), governed by the principle of one person, one vote, which delegated authority to the bodies that guided the cooperative on behalf of the common good.

3. Sovereignty of work. Mondragón declared that it considered work the principal factor to transform nature, society, and individuals. Consequently, the cooperative rejected hiring salaried workers, saw work as essential to produce wealth, and dedicated itself to enabling all its associate workers to have employment in the cooperative.

4. Instrumental and subordinate character of capital. The Mondragón Cooperative Experience considered capital useful but subordinate to work; the disposition of capital was subordinate to the continuity and development of Mondragón; and the use of capital should not impede open membership in the cooperative.

5. Participation in management. The Mondragón Cooperative Experience advocated democracy not only externally in society, but also internally in the cooperative's business operations. Workers should participate in the management of the cooperative. Workers and their representatives should be consulted about the economic, organizational and employment decisions that affect them.

6. Retributive solidarity. The Mondragón Cooperative Experience declared that sufficient and solidary payment to its worker associates was a basic principle of its management. Payment should be sufficient to the extent allowed by the cooperative's resources; and payment should be solidary internally in terms of an appropriate differential payment for work, and externally in terms of payment equivalent to the remuneration received by salaried workers outside the cooperative, as appropriate.

7. Intercooperation. Mondragón advocated cooperation with other cooperatives and related groups, in the Basque country and globally.

8. Social transformation. The Mondragón Cooperative Experi-

ence proclaimed its intention of promoting solidary social transformation in an effort that would promote economic and social reconstruction and build a more full, just, and solidary Basque society. The cooperative would reinvest a majority of its net surplus to create new positions, support community development, provide social security for its workers, and promote the Basque language (Euskara) and Basque culture.

9. Universal character. Mondragón proclaimed its solidarity with all those who work for economic democracy and advocate the peace, justice, and development that are proper to international cooperativism.

10. Education. The Mondragón Cooperative Experience declared that it would dedicate sufficient human and economic forces to cooperative, professional, and youth education.

Mondragón has great diversity in the products manufactured by its distinct member cooperatives. The Ederlan Cooperative (Guipúzcoa Province), with 700 members, does iron and aluminum casting (automobile and computer parts) and machine casting (appliances); Fagor Industrial (Guipúzcoa), 300 members, manufactures ranges, fryers, convection ovens, washing machines, and other household appliances; Guria Industries (Guipúzcoa), 110 members, does shipbuilding and ship repairs; Osatu (Vizcaya Province), with 18 members, makes cardiac monitors and defibrillators; Urssa (Alava Province), 250 members, does boilermaking and steel structures; Ian (Navarra Province), 45 members, produces canned and bottled fruits and vegetables; Behi-Alde (Alava), 26 members, is involved in milk production and the sale of cattle for meat and breeding; and Eroski (Vizcaya), 1,379 members, operates retail stores in the four Basque provinces.

Mondragón over the years has been a source of meaningful work and of worker independence and pride in the Basque country. It has bolstered local economies and benefited local communities. In addition to meeting their own individual needs, and the capitalization and recapitalization requirements of their member cooperatives, the workers of Mondragón have provided substantial funds for community revitalization.

The Mondragón earnings each year, in fact, are distributed to assure cooperative continuity, worker well-being, and societal sustainability. After the workers' remuneration and other expenses are met, at least 20 percent of the surplus is placed in a reserve fund to meet the member cooperatives' needs; about 70 percent is disbursed to the worker associates' personal internal capital accounts (whose interest can be withdrawn, whose balance can be used as collateral for loans taken out by the worker, and whose principal is untouched until the worker retires or leaves the cooperative system), a distribution that is based on the worker's salary grade and number of hours worked; and at least 10 percent is allocated to community needs.

Through recessions in Spain and fluctuating international markets, the Mondragón Cooperative Experience has survived and prospered. Its product diversity and financial stability have enabled it to support financially strapped cooperatives with prosperous cooperatives' earnings; its commitment to its individual worker associates has meant that its members have almost guaranteed job stability: if a cooperative changes managers, or needs fewer workers, the displaced employees receive job retraining as needed, and are shifted to different positions in the same cooperative or in another cooperative.

Mondragón's success as a worker-owned and worker-managed democratically run economic enterprise has made it a model for other cooperatives throughout the world.

Analysis and Projection

Analyze the successes of the Mondragón cooperative in (1) meeting the business needs of its individual members and individual cooperatives; and (2) meeting the social needs of Basque society.

How might the Mondragón experience be replicated in other countries and in similar and different industries?

CHAPTER 4: CONTROL DATA CORPORATION

In 1957 William C. Norris and eight associates founded the Control Data Corporation in Minneapolis, Minnesota. Over the years, CDC became noted for its unique combination of business innovations, social consciousness, and financial success. The corporate slogan under Norris's leadership was "Addressing society's unmet needs as profitable business opportunities in cooperation with the government and other sectors."

Control Data was initially financed by 600,000 shares of stock sold at $1 per share; no one among the 300 shareholders was allowed a controlling share in this first publicly financed computer company. Its initial work took place in an old warehouse in Minneapolis, with initial products being high-end, large-scale computers for scientific and engineering applications. Initial clients were the Atomic Energy Commission, the Defense Department, and large universities: Illinois, Wisconsin, and Michigan State. From these simple beginnings CDC experienced phenomenal growth toward its peak year of 1981, when with assets of more than $5 billion it was ranked high in the Fortune 500, and had thousands of employees in plants and centers throughout this country and in several foreign countries. It became known as one of the 100 best corporations to work for in America.

CDC Programs

Under the leadership of business, technology, and social visionary William Norris and his creative and committed managers, Control Data initiated and/or implemented a unique variety of business innovations and practices. In addition to its successes in computer design and production, CDC developed or integrated into its operations complementary services and products such as the Commercial Credit Corporation financial services; PLATO computer-based educational system and learning programs; CYBERNET data services to benefit businesses that could not afford their own expensive computer systems; CYBERSEARCH, an electronic employment service; Business Advisors, Inc. (BAI), providing high-quality consulting services for small businesses; Control Data Learning Centers for educational programs; Control Data Business and Technology Centers, providing low-cost, flexible space and support services for small companies; City Venture Corporation, to promote urban revitalization through a consortium of business, professional, and religious groups; Rural Venture Corporation, another consortium of business and religious organizations, to revitalize small-scale agriculture by providing a data base, access to PLATO courses, and appropriate technologies through Agriculture, Business and Service Centers (ABCs); Princeton Small Farms Project to enable beginning farmers to initiate small-scale operations using computers for agricultural courses, market projections, accounting, and so forth; Rural Technology Consortium; Microelectronics and Computer Technology Corporation (MCC), a consortium based in Austin, Texas, of more than twenty companies involved in high-tech research and development; Local Government Information Network (LOGIN); Stay Well program to improve employee health; Employee Advisory Resource program (EAR), providing company-wide counseling services for pressing personal, home, and job-related problems; Health Evaluation through Local Processing (HELP), which integrates patient, laboratory, and hospital management; prison education; and the Fair Break program to convert the hard-core unemployed into productive workers.

CDC was known for innovation from its inception. It pioneered much of the early computer research and manufactured the world's fastest computers. It created programs for employee development and interaction—through internal interaction and communication, and personal and professional counseling—that became models for other businesses and were marketed to them.

Control Data integrated its business skill and social philosophy in practical production efforts to revitalize urban and rural areas, alleviate poverty, lower unemployment, and fight racism. Production facilities were located in depressed areas such as Toledo, Ohio; Baltimore, Maryland; Washington, D.C.; San Antonio, Texas; Campton, Kentucky; Minneapolis, Minnesota; and St. Paul, Minnesota. In the latter two locations, CDC established its Northside plant in Minneapolis and Selby-Dale plant in St. Paul to promote urban revitalization through employment and practical needs assistance for the poor of the area (such as day-care services, health programs, and legal services). When conservative business analysts took CDC and William Norris to task for being too social-program oriented, Norris responded that he was not developing social programs, he was using sound business principles and practices to respond to societal needs and making a profit in the process.

Control Data received contrasting reviews from different members of the business establishment and social analysts. For some, Norris and CDC had lost sight of the "fact" that "the business of business is business"; CDC was too involved in "social programs" and not concerned enough about profits and expansion. For others, CDC showed "the human face of capitalism." And for still others, capitalism has no human face, and CDC showed only that there can be alternatives within the capitalist economic structure that are exceptions to the rule, when they are innovative enterprises that have a board and managers committed to social responsibility.

What emerges from all of these assessments is the acknowledgment that Control Data was doing innovative work in the realms of technological development, business structure, and social commitment. This innovation inspired its competitors and other companies to evaluate and alter their own internal policies and external practices.

CDC had a lenient and even encouraging policy toward employees who had an innovative idea and wanted to begin a company of their own. Some of these new ventures became subsidiaries of CDC; others became independent in their own right; still others eventually became competitors. Although CDC by 1993 was much downsized since its heyday, it had given birth to numerous other private-sector ventures, and since small businesses provide most new jobs and most innovation in business, CDC's influence and positive impacts will continue well beyond its own peak years.

As the ripples of its innovations and philosophy continue to extend outward, Control Data is probably having more of an impact on business and technology than would have been possible had it become a tightly controlled transnational corporation; and it will continue to have far more national and global influence in the business world than its competitors, in terms of products, spinoff enterprises, and employee development.

The influence of Control Data Corporation continues today. Its technological innovations are the basis for advances in the computer industry; its employee programs have been adopted by other corporations; and the social involvement it pioneered has inspired other owners and executives to guide their corporations into the realm of responsible interaction with the communities of which they are a part.

Analysis and Projection

Analyze the success of the Control Data Corporation in (1) meeting external social needs through business and technological innovation and community consociation; and (2) forming a highly productive internal work team that (*a*) innovatively responded to business needs through conceptualization and concretization; and (*b*) creatively addressed staff needs through interpersonal and interdepartmental communication.

How might the successful CDC structure and practices serve as a model for the development of new corporations or be used in renovating existing corporations?

Compare Control Data Corporation and Mondragón. What similarities and differences exist? Is there a third model that might be developed incorporating elements of CDC and Mondragón, or are their models appropriate for their differing environments? Other models have been devised for technological development and employment promotion, which will be seen later: a model on a more local scale, with community-based analysis and development (Center for Maximum Potential Building Systems), and a model on a more national scale, with public–private partnerships as its base (Job Creation Collaborative). All of these models are complementary and compatible, when proposed and practiced with social and environmental needs in mind.

CHAPTER 5: CENTER FOR MAXIMUM POTENTIAL BUILDING SYSTEMS

The Center for Maximum Potential Building Systems (CMPBS) is an organization founded in Austin, Texas, in 1974 to develop appropriate technology. Its co-founder and driving force has been Pliny Fisk III, an innovative architect and designer who has traveled the world studying and promoting sustainable housing, energy, and employment, all within the context of sustaining the environment. The center and Pliny Fisk have received numerous awards for their creative work, including recognition for the City of Austin Green Builder Program, which was the sole U.S. recipient of a Local Government Honours Programme award at the United Nations Earth Summit, 1992; the National Center for Appropriate Technology's 1991 Distinguished Appropriate Technology Award in the area of environmental protection; First Place for design in the Multi-Family division, in the National Endowment for the Arts "Designing for Area Resource Efficiency" International Competition, 1986; and selection in 1981 by the U.S. Department of Housing and Urban Development as one of ten outstanding community-based renewable energy organizations in the United States from among 600 reviewed.

The Center for Maximum Potential Building Systems focuses on working with individuals and communities to create housing and sustainable economic development projects that utilize local resources to the greatest extent possible. The center strives to maintain a balance between technical sophistication and local access to resource technologies; projects are tailored to and by local communities to fit the local resource base and the traditions, aptitudes, and aspirations of the local user population. Training and educational programs are provided to ensure the long-term viability of projects developed with local and regional communities. All projects strive to utilize appropriate technologies and regional resources while balancing human needs with environmental considerations.

CMPBS Projects

On September 23, 1977, the LoVaca Gathering Company cut off the natural gas supply to Crystal City, Texas, in a dispute over a rate increase. As winter approached, the residents of this "Spinach Capital of the World," an impoverished community comprised predominantly of Mexican American citizens, wondered how they might meet their energy needs for home heating, hot water heating, and cooking. City leaders approached Pliny Fisk for assistance, and CMPBS began to investigate means of meeting the city's heating and cooking needs.

Community members working with CMPBS discovered surplus Korean War stoves in nearby San Antonio that could serve for heating homes and cooking meals if sufficient wood fuel could be obtained. Further research revealed the availability of mesquite wood, which burns well and long, in the river bottoms of the surrounding area. Mesquite was gathered in trucks by community teams.

With home heating and cooking needs met, Fisk turned to the problem of providing energy for hot water heaters. Using innovative thinking, he designed a new type of solar collector constructed with recycled materials—used fluorescent tubes from local public schools and used printing plates from an area publisher. The solar collector, resembling an old Gatling gun with its circle of tubes on the black-painted printing plates, not only provided the energy for the hot water system but also created a new cottage industry in Crystal City for its manufacture.

The CMPBS project in this case met a pressing community need; used a renewable energy source (solar); recycled materials that otherwise would have been wasted as "trash"; and created a new source of employment. In the process, a local community learned new skills, new uses of renewable resources (mesquite trees and sunlight), new uses for old products (surplus stoves, fluorescent bulbs, and printing plates), and a new self-sufficiency.

The most recent and most highly sophisticated CMPBS project was developed in 1991: the Biom-metric™ sustainable design methodology, a system of planning and design that represents resources, processes, and products in computer-manipulated icons. The Biom-metric™ program employs in part the UNESCO system of global planning that uses biomes as the basis for information sharing among people engaged in efforts at sustainable development. (Biomes are areas in different parts of the world that have similar flora and fauna, and similar patterns in soils, climate, and hydrology. There are fourteen basic biomes on the earth.) At the biome level, the condition of the air, land, water, and all life forms indicates areas needing enhancement or correction to promote the sustainable interaction of these components in the ecological area. In the Biom-metric™ system, a life cycle assessment (LCA) is made of the components of the biome from "source to sink," that is, from the initial condition of the object studied to its final disposition, and through all stages in between. During the LCA, integration analysis connects the processes of a biome into continuous flows and links them to one another in functionary roles.

When a community, or a segment of it, or planners within it linked to prospective developers or users of its material, energy, money, information, or human resources, engages in the LCA process and represents its findings with icons, plans for sustainable development can be formulated through the manipulation of icons on computer screen maps before any real-world changes are made to the biome locally or regionally. In the mapping process, people become aware of the resources of their area, of the sources of the products used in such processes as home construction and food consumption, and of potential sources of alternative products or byproducts that might be available locally to meet their needs. In sustainable development planning, the icons can be manipulated to change input materials and output products: a community can choose to import alternative materials to meet its residential and infrastructural needs and to export its "waste" materials in alternative ways as usable products: for example, sewage converted to compost or sawdust converted to insulation bricks, or sulfur converted to building blocks.

In the area of input sources, a community importing portland cement for construction projects might find, upon analysis of its local resources, that it has available an appropriate type of ash from volcanic activity or from a coal-fired power plant that can be used as a replacement cement. By developing this resource (which might be seen as merely a "waste" byproduct in the case of the coal-fired plant), the community can provide jobs for its citizens both directly (workers at the cement production facility) and indirectly (workers at local businesses in which the cement production workers will shop); lessen construction costs for local contractors and their residential, commercial, and industrial clients; diminish the use of energy for portland cement production and transportation; and reduce the community need to find a site for disposal of a "waste" product (the ash from the coal-fired plant). Fisk and the center have developed a third use for the initial resource of coal, in addition to its uses as fuel and as cement: sulfur is extracted from the coal to manufacture building blocks. Thus, a resource that in many areas serves only one purpose, energy production (and causes pollution of the air and eventually the land and water through its smokestacks), can be used in its entirety: as a fuel, as a cement, and as a building component. It can also provide diverse kinds of jobs, lessen regional energy consumption, and diminish or eliminate sulfur-based pollution, such as acid rain, thereby enhancing local and distant environments.

In the area of output products, a community might evaluate its wastewater disposal, and conclude that it might recycle some of the water after treatment and use some of the sludge for compost for area farmers and gardeners. In this case, water consumption would be reduced, new jobs would be provided, agriculturalists would have a local source of organic fertilizer, and regional water quality could be enhanced. Similarly, a timber-based community could develop new uses for such "waste" byproducts as sawdust: for stove fuel as compressed pellets, for window or door frames when fused with plastic, or for building insulation as foamed bricks.

The Center has begun working in collaborative programs with educational institutions, government agencies, community leaders, architects, engineers, business leaders, and ordinary citizens to promote appropriate technologies and sustainable economic development using the Biom-metric™ system. As members of one biome develop ideas, processes, and products, their developments can be shared with people in similar biomes across the globe; and as each type of biome progresses on the path to sustainable development, the earth as a whole benefits.

Analysis and Projection

Analyze the activities of the Center for Maximum Potential Building Systems in terms of their potential for (1) providing a viable basis for meeting human needs; (2) promoting sustainable development; (3) providing good jobs; and (4) protecting the environment.

Is the CMPBS methodology applicable on a global scale? What impediments might there be to sustainable development based on the CMPBS model? How might these obstacles be overcome?

Compare the Mondragón, Control Data Corporation, and Center for Maximum Potential Building Systems approaches to resource utilization and employment provision. What aspects of each might be appropriate in specific local and regional contexts? How might the experiences of each inform the practices of the others?

CHAPTER 6: JOB CREATION COLLABORATIVE

In 1993, the William C. Norris Institute of Minneapolis, Minnesota, developed and began to implement the Job Creation Collaborative (JCC), an enterprise based on public–private cooperation to better utilize technology in the creation of good jobs offering a living wage. Intended to develop technology-oriented jobs with the help of the combined resources of entities of the public and private sectors, the JCC was an outgrowth of efforts made by William C. Norris, while he was CEO of the Control Data Corporation (CDC), to promote community development using corporate resources in conjunction with government commitments.

Plans to Surmount Difficulties

The original CDC effort had encountered several difficulties: the limited resources available even to a large corporation interested in meeting society's unmet needs; adverse criticism of CDC efforts by some stockholders and security analysts, which hindered implementation; lack of full community consensus for CDC proposals, linked in part to community members' fears of the motives of CDC (for example, that the corporation was seeking to take advantage of poor people); failure of government entities to meet their commitments; and lack of public understanding of the role of the technological innovation process in creating good jobs.

The Job Creation Collaborative is structured to surmount these difficulties and to benefit from the experience and practices of other programs. Since the Norris Institute is a nonprofit corporation, community fears about corporate domination are allayed and individual corporate participants in the program are isolated to some extent from the risk of failure. Since profits generated are reinvested in the job creation effort, communities understand that no corporation is seeking to take advantage of poor people. Since participating cities and counties must make a twenty-five-year political and financial commitment to the collaborative, the problem of local entities not fulfilling their responsibilities to the effort is eliminated.

A basic premise of the JCC is that the source of most good jobs is innovation: applying technology to create new and better products or new and better uses for existing products or byproducts. The expansion of the innovation process requires the conversion of research outcomes into commercial products. At present, small companies lack the financial resources, personnel and physical plant needed for extended research; because of their small size and concomitant risk to potential investors, they also lack access to capital to implement completed research in a production process. At the same time, large corporations, educational institutions, and government laboratories have greater resources for extended research and technology development than small companies but convert only a small percentage of their findings into products and services because of increasing costs to commercialize research; increasing concerns of corporations for short-term profits and of government entities for accountability for expenditures of tax revenues; and rapid technological advances that render some products obsolete shortly after their production or distribution.

The Job Creation Collaborative distributes risks and benefits among the public and private partners involved in its program. It also provides key types of assistance for technological innovation: early-stage capital to assess product feasibility, develop prototypes, and evaluate market potential; expert assistance—mentoring and monitoring—to help prevent technical and management problems and resolve them expeditiously when they do occur; access to technology for new and emerging companies, and among the collaborators; joint ventures to promote international market opportunities; adaptive organizational structures that relate to changing conditions; linkages between educational institutions and companies that promote research sharing and job development; and technology-based education courses about the requirements for creating good jobs, viable company operations, and sustainable community development.

The JCC assistance process begins with the submission to the JCC of a proposal for a new company, based on research results concerning technological development of a product that has commercial potential. Business and technical experts evaluate the proposal in terms of its financial and technical feasibility. If a positive evaluation emerges from both kinds of experts, financing is provided by the early-stage fund to commence product development. In the development stage, monitoring and mentoring are ongoing, and financing is continuously provided, contingent upon satisfactory progress (as determined by the experts involved on behalf of the JCC), until a demonstrable prototype is completed. Other investors are then invited to participate in the company.

Different organizations have complementary collaborative roles to play in the JCC program. Local counties and cities provide capital for early-stage financing and for administrative costs of the JCC, receiving in return the location of new companies in their area. (New companies will repay this investment to the JCC, to be reinvested in the community source of their funds, when they reach a viable level of operation, or when they go public or are acquired by another firm.) Universities and colleges provide researchers and laboratory facilities and, in turn, will see their research concretized into production and will subsequently receive funding for further research.

Large companies assist in the spin-off of technologies they develop when they are not pursuing commercialization, invest in start-up companies developing a product in a field of their interest, and collaborate in the manufacturing and/or marketing of start-up company products. U.S. government laboratories commercialize research results through start-up companies funded by the JCC, thereby utilizing more effectively the taxpayer dollars funding their research. Foundations interested in promoting community economic development make grants to selected communities and to the JCC for the dissemination of education courses on technology development and benefits, and for the development of knowledge-based management assistance tools. Labor unions build closer relationships between educational institutions, companies, and individual workers, including assistance with worker understanding of and participation in lifelong learning and job creation processes. Finally, community organizations provide knowledge of local cultures, needs, and concerns; human support services; identification of appropriate community residents for education and employment; assistance in training; and promotion and delivery of computer-based education courses on the process of job creation and its importance for community development and stability.

The Job Creation Collaborative already involves several Minnesota companies and communities, and has linked them with University of Minnesota research facilities. The William C. Norris Institute thus has initiated a creative and long-term public–private partnership dedicated to responsible community development through the creation of technology-based jobs. The program developed by the Norris Institute could be a model for public–private partnerships designed to allow companies to compete more effectively in the international arena as well, where some nations or consortia of nations already utilize such partnerships.

Analysis and Projection
Evaluate the potential for public–private partnerships to effectively utilize resources available to the public and private sectors for promoting technological innovation. How might such partnerships be made workable?

Are public–private partnerships becoming essential for nations lacking them, because of the international economic power exercised by countries where they already exist?

In what other areas might public–private collaborative efforts benefit regional and national communities, including a linkage of sustainable economic development and environmental sustainability?

Analyze the salient characteristics of the organizations described in the four case studies: the Mondragón Cooperative, Control Data Corporation, the Center for Maximum Potential Building Systems, and the Job Creation Collaborative. How do their respective theories and practices complement or contradict each other? How might they be made more complementary?

Project the applicability of these models in different types of social contexts, in the present and in the future.

APPENDIX B

Reflection Questions

Ideas presented in *Ethics and Technology* serve as the basis for reflecting on present and projected technological practices in community and environmental contexts.

CHAPTER 1

1. Why might someone developing a potentially socially harmful product claim to be "value neutral" in that development process?
2. In which resource areas today might continued "growth" be detrimental in the long term? What alternatives would you suggest?
3. How would you resolve a conflict between two industries vying for the same resource: a logging company and an outdoor sporting company each seeking to use a particular segment of public land for their operations? Each claims that a "multiple use" policy gives them the right to use that forest; each claims that it will provide needed jobs for the region; each claims that it would provide a needed economic boost to the region. What values or principles might you see operative in this situation? How would you resolve a similar conflict between a cattle firm that wants to graze livestock in a wilderness area and an energy firm that wants to strip-mine coal in the same area? What values or principles might you see operative in this situation?
4. What societal need might be addressed today as a business or scientific opportunity?
5. How might nations work together on responsible economic development?
6. How might the ethics of transformation guide technology enterprises to develop processes and products beneficial to society?

CHAPTER 2

1. What individual interests or rights should be subordinate to societal interests or rights, international interests or rights, or environmental interests or rights?

2. When should those individual interests or rights be subordinate to societal, international, or environmental interests or rights?

3. Is it possible to be a virtuous businessperson in an immoral business milieu?

4. Should people expect less of their society or their corporations than they expect of individuals?

5. Reflect on the fish story. How might people's needs be met both today and tomorrow?

6. Discuss the ethical principles. For what carefully considered reasons do you agree or disagree with them?

CHAPTER 3

1. How might managers both promote subsidiarity in operations and accept responsibility for results?

2. How might owners encourage employee creativity to devise operational and product improvements?

3. How might workers be both responsible citizens and loyal employees if they believe a company product to be harmful to consumers or to the environment?

4. Is the creation of cooperative enterprises desirable or even possible in a variety of types of national settings? What might be some impediments to the formation of cooperative enterprises? What might be some social stimuli for the promotion and establishment of cooperative enterprises?

5. Discuss the ethical principles. For what carefully considered reasons do you agree or disagree with them?

CHAPTER 4

1. In what ways might technological advances be harmful to the community in which they are operative or to the broader global community?

2. What responsibilities to local communities should be fulfilled by corporations engaged in extracting natural resources, formulating potentially polluting productive processes, or releasing harmful products into the air, land, or water?

3. What local benefits should or might a company provide in addition to jobs?

4. How should a business exercise responsibility in a negative con-

text, that is, to correct a harm the business has caused to a community?

    5. How should a business exercise responsibility in a positive context, that is, to remedy a community need?

    6. Analyze one of the environmental disasters cited in the "Community Relations" section. How might the corporate or governmental entity involved have acted more responsibly toward the community and toward the environment?

    7. Trace an innovative idea—an original one or a previously concretized one—through the seven steps to innovative development.

    8. Discuss the ethical principles. For what carefully considered reasons do you agree or disagree with them?

CHAPTER 5

    1. How might it be possible to have both employment and environment in such areas as logging, mining, farming, manufacturing, and energy generation?

    2. What recyclable resources or alternative extractive or productive processes might benefit manufacturing plants, society, and the environment?

    3. Would regional or cooperative ownership and development of some resources be more socially advantageous than corporate or national ownership?

    4. Should animals have natural rights? Should natural rights be extended to plant life? air? land? water? rocks?

    5. Discuss the implications for economic development—potential positive and negative impacts—if natural rights were to be extended to include more than human beings.

    6. Why should indigenous peoples' right to control their lands and destiny be respected? If they are not, can any group hope to retain its own rights in the face of efforts by more-powerful groups to control its future prospects?

    7. How might the seven "Steps for Appropriate Technological Development" assist people's efforts to promote sustainable economic development, sustainable employment, and sustainable environmental integrity?

    8. Discuss the ethical principles. For what carefully considered reasons do you agree or disagree with them?

CHAPTER 6

1. Is enlightened self-interest a sufficient basis for promoting responsible technological development, management, and utilization?

2. What attitudes toward money and toward the environment drawn from spiritual traditions might promote both economic development and ecological responsibility?

3. What individual, social, and environmental benefits might be fostered through personal or societal relations to spiritual realities?

4. How might spiritual insights and scientific insights be integrated into a holistic effort to promote care for the earth and responsible technological development?

5. What ethical principles might you derive based on integration of spiritual insights and scientific insights?

6. Discuss the ethical principles. For what carefully considered reasons do you agree or disagree with them?

CHAPTER 7

1. What barriers to social, technological, and spiritual progress are raised when beneficiaries of the status quo define "reality," "realistic," and "utopian"?

2. What values of and attitudes toward ethical technological development should be part of business plans and practices?

3. What utopian ideas and projects should business and technology implement in the next decade?

4. Review the four case studies: Mondragón, the Control Data Corporation, Center for Maximum Potential Building Systems, and the Job Creation Collaborative. How might any or all of these serve as models to promote a positive and productive relationship among ethics, environment, entrepreneurship, employment, and economics in technological innovation? What aspects of these models might be used in programs or projects of which you are aware today? What do you envision might be the future uses of aspects of these models?

APPENDIX C

Projects

PROJECT 1 (CHAPTER 3)
Design a process for promoting both the company good and the common good in an area or around an issue where currently company and community are in conflict, or where technological innovation has the potential for promoting such conflict. The process should include consideration of how the internal philosophy and workings of the company foster or could foster concern for the community at all levels of competency; and how the dynamics of the community and ethical considerations accepted by company and community might be integrated in cooperative efforts to promote the well-being of both company and community.

PROJECT 2 (CHAPTER 5)
Analyze a school or business environment to evaluate areas of resource waste. Is paper (from memos, letters, tests, reports, discarded mail, and so forth) being recycled? What about plastic and glass? Can leftover cafeteria food be donated to needy people (when appropriate and healthy) or used or sold for compost or animal consumption? What happens to aluminum cans from beverage machines: do employers and employees in a business environment, or students and faculty in a school environment, strive to recycle them? Are energy-saving devices and practices in operation? What recycling effort is most imperative?

Devise a recycling program for the most-needed area. Talk to colleagues or to students (perhaps a particular organization in search of a meaningful project and/or a fund-raiser), to managers or faculty, and to owners or administrators, about ways to implement your program so that it becomes an integral part of company or campus life, benefiting people and planet. Explore cooperative efforts with community organizations (environment-oriented, youth-based, senior-based, service); businesses (particularly those selling recyclable products: determine their objections/concerns as well as structures and practices that could be acceptable to them); and area schools. Discuss government participation in recycling efforts with appropriate staff and elected officials of your community, county and state, as appropriate and helpful to an extended and ongoing recycling program.

PROJECT 3 (CHAPTER 7)

Develop a plan for a project to use an appropriate, earth-respective technology to meet a social need in an economically viable, employment-generating, and environmentally sustainable enterprise (a new enterprise or a new component for an existing enterprise). The project might be researched and developed in collaboration with local businesspeople, government staff, nonprofit organization staff, other students, and interested citizens (including potential entrepreneurs and employees).

In your rationale for your project, include a statement of the social need; ethical considerations in striving to meet that need; and your reasons for selecting the particular technology you advocate.

Predict the long-term effects of your project on the local community (in terms of job promotion, economic vitality, and fulfillment of the objective of meeting a social need), and on the regional environment (use of resources; negative and positive impacts).

Append to your plan comments from people who have reviewed your plan and would like to see it implemented.

SELECTED BIBLIOGRAPHY

Barbour, Ian. *Ethics in an Age of Technology.* San Francisco: HarperCollins, 1993.
Berry, Wendell. *The Unsettling of America: Culture and Agriculture.* New York: Avon Books, 1978.
Boff, Leonardo. *Ecology and Liberation: A New Paradigm.* Maryknoll, N.Y.: Orbis Books, 1995.
Brueggemann, Walter. *The Land.* Minneapolis: Fortress Press, 1977.
Cobb, John B., Jr. *Sustainability: Economics, Ecology, and Justice.* Maryknoll, N.Y.: Orbis Books, 1992.
Fedorov, E. *Man and Nature: The Ecological Crisis and Social Progress.* New York: International Publishers, 1980.
Fisk, Pliny, III. "Towards a Theory and Practice of Sustainable Design." Presentation to the 1992 National Convention of the American Institute of Architects. Austin, Tex.: Center for Maximum Potential Building Systems, 1992.
Hart, John. *The Spirit of the Earth: A Theology of the Land.* Ramsey, N.J.: Paulist Press, 1984.
Jackson, Wes. *Altars of Unhewn Stone: Science and the Earth.* San Francisco: North Point Press, 1987.
———. *Becoming Native to This Place.* Lexington: University Press of Kentucky, 1994.
Leopold, Aldo. *A Sand County Almanac.* New York: Oxford University Press, 1987.
Levering, Robert, Molton Moslowitz, and Michael Katz. *The 100 Best Companies to Work for in America.* New York: Signet Books, 1987.
Liebig, James E. *Business Ethics: Profiles in Civic Virtue.* Golden, Colo.: Fulcrum Publishing, 1990.
Lovelock, James. *The Ages of Gaia: A Biography of Our Living Earth.* New York: W. W. Norton, 1988.
Lyons, Oren, and John Mohawk, eds. *Exiled in the Land of the Free: Democracy, Indian Nations, and the U.S. Constitution.* Santa Fe: Clear Light Publishers, 1992.
Medvedev, Zhores A. *Nuclear Disaster in the Urals.* New York: Vintage Books, 1980.
Moltmann, Jürgen. *God in Creation.* New York: HarperCollins, 1991.
Nash, Roderick Frazier. *The Rights of Nature: A History of Environmental Ethics.* Madison: University of Wisconsin Press, 1989.
Neihardt, John G., ed. *Black Elk Speaks.* New York: Pocket Books, 1972.

Norris, William C. *New Frontiers for Business Leadership*. Minneapolis: Dorn Books, 1983.

Ormaechea, Jose María. *La Experiencia Cooperativa de Mondragón*. Mondragón, Basque Region, Spain: Otalora, 1991.

Orr, David W. *Ecological Literacy: Education and the Transition to a Postmodern World*. Albany: State University of New York Press, 1992.

———. *Earth in Mind: On Education, Environment, and the Human Prospect*. Washington, D.C.: Island Press, 1994.

Price, Robert M. "Commitment without involvement." *Harvard Business Review* (Sept./Oct. 1984), 162.

The Holy Qur'an. 3d ed. Translated and with commentary by Abdullah Yusuf Ali. New York: Tahrike Tarsile Qur'an, Inc., 1987.

Rasmussen, Larry L. *Earth Community, Earth Ethics*. Maryknoll, N.Y.: Orbis Books, 1996.

Ruether, Rosemary Radford. *Gaia and God: An Ecofeminist Theology of Earth Healing*. New York: HarperCollins, 1994.

Schumacher, E. F. *Small Is Beautiful: Economics As If People Mattered*. New York: Harper & Row, 1973.

Schumacher, E. F., with Peter F. Gillingham. *Good Work*. New York: Harper & Row, 1979.

Shannon, Thomas. *An Introduction to Bioethics*. 3d ed. Ramsey, N.J.: Paulist Press, 1997.

Standing Bear, Luther. *Land of the Spotted Eagle*. Lincoln: University of Nebraska Press, 1978.

Swimme, Brian, and Thomas Berry. *The Universe Story: From the Primordial Flaring Forth to the Ecozoic Era*. San Francisco: HarperSanFrancisco, 1992.

Thomas Aquinas. *Summa Theologica*. Vol. 2. Trans. Fathers of the English Dominican Province. New York: Benziger Brothers, 1947.

Tuleja, Tad. *Beyond the Bottom Line: How Business Leaders Are Turning Principles into Profits*. New York: Penguin Books, 1987.

Whyte, William Foote, and Kathleen King Whyte. *Making Mondragón: The Growth and Dynamics of the Worker Cooperative Complex*. Ithaca, N.Y.: ILR Press, Cornell University, 1988.

Wilson, Edward O. *Biophilia*. Cambridge: Harvard University Press, 1984.

———. *The Diversity of Life*. Cambridge: Belknap Press of Harvard University Press, 1992.

Worthy, James C. *William C. Norris: Portrait of a Maverick*. Cambridge, Mass.: Ballinger Publishing Company, 1987.

PRAISE FOR
ETHICS AND TECHNOLOGY

"John Hart, one of a new group of ethicists to whom we all have to listen, argues persuasively that we need to do our homework, not only in how technology works in the modern world, but in how ethics 'works' as well. We are never asked to compromise either the life of reason or the life of commitment."
— Robert McAfee Brown, Professor Emeritus of
Theology and Ethics, Pacific School of Religion

"*Ethics and Technology* blends theoretical exploration of ethical issues with real-life situations and decisionmaking. The book is very 'user friendly' and badly needed. I highly recommend it."
— Elisabeth Schüssler Fiorenza, Krister Stendahl Professor of
Scripture and Interpretation, Harvard Divinity School

"Religion, the seedbed of morality, holds the keys for the expansion of ethics to include nature. John Hart is a pioneer in environmental theology, and his book promises to open the ethical doors leading to a full sense of community on Planet Earth. If, as a species, we can recognize the wisdom in Dr. Hart's message, we have a fighting chance to sustain ourselves and our environment into the twenty-first century and beyond."
— Roderick Frazier Nash, Professor Emeritus,
University of California, Santa Barbara

"*Ethics and Technology* addresses some of the most urgent issues facing American business in light of the importance of the common good. It will stimulate practical ethical reflection on cutting-edge questions by both business leaders and citizens."
— David Hollenbach, S.J., Margaret O'Brien Flatley
Professor of Catholic Theology, Boston College

"For anyone whose church or other spiritual influence seems disappointingly inadequate for guiding human behavior in our materialis-

tic world generated by ever-larger flows of products and services, John Hart's seminal work *Ethics and Technology* may provide the reader, if he or she is determined, not a wholly new but a refocused, more detailed, more pragmatic ethic, with powerful possibilities. The author builds a meticulous, extensive, and, enlivened by his frequent illuminating, provocative 'case histories,' scholarly, precisely defining exposition of ethics and, particularly, extensions thereof to the world of technology and expanding human relationships—individual, group, community, city, state, world. Perhaps the most important contribution of this work is its extension beyond those human interrelationships to the planet earth on which we live, warning that absent its inclusion, we won't! For utopia seekers, a new world; for the concerned, a challenge!"
  —Judson Bemis, Retired CEO, Bemis Company, Inc.

"How is technology related to ethics, the individual to society, employer to employee, corporation to community, entrepreneurship to environment, spirit to science? These are some of the critical questions which John Hart explores in his interactive study *Ethics and Technology*. This will be a helpful resource for people in societies around the world who are struggling with the conflicts and possibilities of technological systems at the dawn of the twenty-first century."
  —Rosemary Radford Ruether, Georgia Harkness Professor of
  Applied Theology, Garrett-Evangelical Theological Seminary

"John Hart offers an impressive balance between realism and idealism—realism in discussing technology in the context of real-life situations and institutional constraints, idealism in affirming that we do have opportunities, as individuals, communities, and businesses, to redirect technology towards the fulfillment of human life and the preservation of our environment."
  —Ian G. Barbour, author of *Ethics in an Age of Technology*,
              and Professor Emeritus, Carleton College

"*Ethics and Technology* helps the reader recognize and tackle the ethical challenges faced by anyone working on the cutting edge of technology today. John Hart believes that 'the union of ethics and technology can transform the world.' Hart guides the reader to an understanding that religious and ethical traditions are not counter to

scientific progress, but rather can (and must) be intelligently used to guide appropriate conduct in the present, setting the stage for a better future for everyone."
—Ron Kroese, Executive Director,
National Center for Appropriate Technology

"John Hart here proposes for the general reader a clear, very readable, user-friendly, convincing vision of a more holistic understanding of business and technology that challenges the present reality. The book skillfully develops and applies a communitarian rather than an individualistic ethic, which entails significant ramifications for our society. I hope that many people will not only read this important book but put its transformative ethics into practice."
—Charles E. Curran, Elizabeth Scurlock University Professor
of Human Values, Southern Methodist University

"As concise and lucid a treatment of ethics and technology for practitioners as I have seen. . . . I highly commend this work and the conscientious and helpful spirit in which it is written."
—Larry L. Rasmussen, Reinhold Niebuhr Professor of
Social Ethics, Union Theological Seminary, New York

"*Ethics and Technology* will be a welcome addition to the literature on ethical issues in technology. Of particular importance is the attention given to both individual and social responsibility. Additionally, the material is clearly and thoughtfully presented, the discussion questions are helpful, and the reflection sections help focus the material. This will be a very helpful book for all students."
—Thomas A. Shannon, Professor of Religion and
Social Ethics, Worcester Polytechnic Institute

"John Hart shows convincingly that in this technological age few decisions are solely technical. Human purposes and ethical choices are deeply imbedded in all policies of science, business and industry, economic development, and government. He works persistently to harmonize diverse interests that often conflict. To those who ask whether he is too optimistic, he provides just enough examples of actual success to justify hope."
—Roger L. Shinn, Professor Emeritus,
Union Theological Seminary

"*Ethics and Technology* challenges the reader's thinking with both critical philosophical analysis and practical application. Hart provides the insight, spirit, and ethical analysis that makes a convincing case for social responsibility and communitarian ethics as our passport to a just and sustainable future."

—Pamela Mavrolas, Executive Director,
Alternative Energy Resources Organization

"In an age when technological innovation and development is moving forward at an ever-increasing pace, Professor Hart raises critical issues regarding the impact of technology on working people and society at large. His work underscores the need to go beyond the economic realm when looking at progress and, in fact, to incorporate an entire range of human values in determining the course of future development. At a time when so many people despair about the world of tomorrow, this book reassures us that the future is really ours to create."

—Thomas Donahue, National Secretary-Treasurer,
AFL-CIO

"*Ethics and Technology* brings together for the first time the critical elements that will be required for society as we have known it to survive. Technological change has always been accompanied by problems as well as great promise. This is especially true today as the increasing rate of technological change is accelerating all of us into the next century. John Hart has taken a fresh and long overdue look at the ethical issues and dilemmas that accompany technological change. *Ethics and Technology* is must reading for business, government and citizens.

"Hart clearly defines the roles and responsibilities that citizens, government, and industry leaders and entrepreneurs must assume if we as a society are to succeed. The ethical principles presented promote individual and corporate responsibility for people as individuals and as members of society. The brilliantly constructed case studies help the reader resolve the competing issues of the individual, society as a whole, the business leader, and technologists. *Ethics and Technology* forces readers to consider their role in the world we are creating. A wonderful and holistic approach."

—Mark Ames, Chief Executive Officer,
Applied Quality Systems

"The developments of modern technology cause many agonizing human problems.... It is vitally necessary that such problems be tackled humanely with respect for everyone's rights, and this requires careful study of the ethical principles that should govern wise decisions. A thorough study of such problems, with challenging case histories, is provided by Professor Hart's book. It is a most valuable source for all who are concerned with the problems of modern technology."
— Peter Hodgson, Ph.D., Head, Nuclear Physics Group,
Oxford University, England

"*Ethics and Technology* reached my desk as I am forming a company to utilize new horse-logging technologies for sustainable, selective harvest of hardwoods for a premium, environmentally certified market. It raises the tough questions that must be considered by those who would earn a living without violating the community of life."
— Richard Cartwright Austin, S.T.D.,
Appalachian Ministries Educational Resource Center

"Dr. Hart offers a new and interesting perspective on what is the fastest-growing segment of our economy. Fortunately, most businesspeople agree that strong ethical considerations must be paramount."
— Ed Jasmin, President (Retired), Norwest Bank, Helena

"John Hart blends insight, compassion, vision, and a touch of humor as he makes us think about balancing values, benefits, and the sometimes whimsical consequences of technology. Farmers are constantly gravely affected by technology and its whims. John Hart demonstrates an understanding of and sensitivity to that fact, as he includes us in his examples and in his challenge to responsibly explore tradeoffs in our pursuit of progress."
— Norb Berg, Boyer Creek Ranch, Wisconsin

"*Ethics and Technology* is a lucid and valuable discussion of one of the most important and vexing subjects on the human agenda. John Hart has done an admirable job of making this topic accessible to technologists, students, and the wider public. Highly recommended."
— David W. Orr, Professor and Chair,
Environmental Studies Program, Oberlin College

"I often feel lost in the struggle between science, advanced technology, and the spiritual need 'to till and to keep' creation. John Hart's book . . . is an educational tool for those who wish to preserve the integrity of creation. He shares his profound knowledge of scientific data and technical aspects of environmental issues—so I have learned how to become an informed caretaker of the earth. Hart has long worked with communities concerned with the spiritual theme of the fate of our Mother the earth. The readers of this well-reading book will learn about a badly needed ethics of transformation."

—Dorothee Sölle, Hamburg, Germany,
Theological Writer and Former Professor of Systematics,
Union Theological Seminary, New York

"Here is an ethic for those who care about science, technology, and business—and who are curious about John Hart's perspective that spiritual values, scientific insights, and social concern together can jointly stimulate truly progressive community development that is ecologically sound."

—Carol S. Robb, Margaret Dollar Professor of
Christian Social Ethics, San Francisco Theological Seminary

"Too seldom do we approach our looming environmental issues in ways that bring forth experienced attempts at solutions. Similarly, seldom do we realize the careful fit between environment, technology, and people that must be made. John Hart has done a masterful job in bringing all these issues to the forefront, with *ethics* as the cornerstone. A timely contribution!"

—Pliny Fisk III, Center for Maximum Potential
Building Systems, Austin, Texas

"Dr. Hart's ethical reflection on technology in relation to the common good is greatly needed today. His insights should facilitate a deeper responsibility for the stewardship of technology and care for God's creation."

—Bishop William Skylstad, Diocese of Spokane, Washington